Calculus III
Exam File

D. R. Arterburn, New Mexico Institute of Mining and Technology, Editor; Bill
Bompart, Augusta College; Peter Braunfeld, University of Illinois at Urbana-
Champaign; William E. Demmon, University of Wisconsin Center - Manitowoc;
Constance M. Elson, Ithaca College; Joe Flowers, Northeast Missouri State
University; Michael E. Frantz, University of La Verne; Biswa N. Ghosh, Hud-
son County College; LeRoy P. Hammerstrom, Eastern Nazarene College; John
H. Jenkins, Embry-Riddle Aeronautical University; Ann F. Landry, Dutchess
Community College; David H. Lankford, Bethel College; Eric M. Lederer,
University of Colorado at Denver and Red Rocks Community College; John
Martin, Santa Rosa Junior College; Thomas A. Metzger, University of Pitt-
sburgh; Alejandro Perez, Laredo Junior College; Calvin E. Piston, John Brown
University; John Putz, Alma College; Michael Schneider, Belleville Area Col-
lege; Alan Stickney, Wittenberg University; Joseph F. Stokes, Western Ken-
tucky University; Norman Sweet, State University College; Bill W. Vannatta,
Temple Junior College; Robert P. Webber, Longwood College; Joseph E. Wiest,
West Virginia Wesleyan College

ENGINEERING PRESS, INC.   SAN JOSE, CALIFORNIA

Donald G. Newnan, Ph.D.
EXAM FILE Series Editor

Printed in the United States of America

5 4 3 2 1

Library of Congress Cataloging-in-Publication Data

Calculus III exam file

   (Exam file series)
     1. Calculus--Problems, exercises, etc.   I. Arterburn,
D. R. (David R.),   II. Bompart, Bill.
III. Title: Calculus 3 exam file.   IV. Series.
QA303.C167   1986   515'.076   86-13534
ISBN 0-910554-63-3

Engineering Press, Inc.   **P.O. Box 1**   **San Jose, California 95103-0001**

Contents

4. MULTIPLE INTEGRALS

Foreword

It is common practice on college campuses for student organizations to maintain files of individual professors' past exams and homework assignments. These files have helped to improve the grades of many students. For this book we have solicited exam problems from college professors all over the country, representing different approaches to the calculus topics. We have not attempted to "homogenize" the problems, preferring to leave the individual flavors intact. Each solution has been prepared by the professor who wrote the examination problem.

This volume is the third of three covering a standard college calculus course, approximately one semester of the usual three semester series. The suggested way to use this book is to choose a problem from the area of interest and work it yourself before looking at the professor's handwritten solution. We have attempted to eliminate errors, but if you should discover any, we would appreciate a note sent to Engineering Press.

We hope this book will improve your test scores, and that you will try the other volumes of calculus problems.

D. R. Arterburn
Editor

1

VECTORS IN THE PLANE AND PARAMETRIC EQUATIONS

VECTORS IN THE PLANE

■■■-**1-1**

Given $\vec{u} = 2\vec{i} - \vec{j}$ and $\vec{v} = 3\vec{i} + 2\vec{j}$, find

a) $\vec{u} - \vec{v}$

b) $|\vec{u}|$

****

a) $\vec{u} - \vec{v} = (2\vec{c} - \vec{j}) - (3\vec{c} + 2\vec{j})$

$= 2\vec{c} - \vec{j} - 3\vec{c} - 2\vec{j}$

$= -\vec{c} - 3\vec{j}$

b) $|\vec{u}| = \sqrt{2^2 + (-1)^2}$

$= \sqrt{5}$

1

1-2 ∎∎

In a triangle ABC, let P be a point located two-thirds of the way from the vertex at A to the midpoint of the opposite side BC. Use vector methods to find the coordinates of P in terms of the coordinates of the vertices. Show from the above that the three medians of the triangle intersect at P.

****

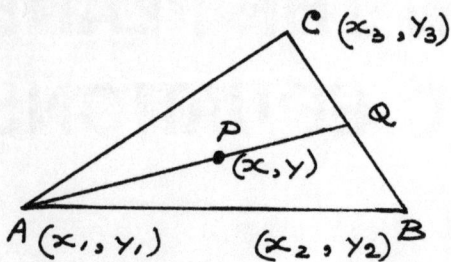

A triangle ABC is shown alongside, with arbitrary coordinates assigned to each vertex. From the problem statement we have Q as the midpoint of B&C.

Hence, $Q = \left[\frac{1}{2}(x_2 + x_3), \frac{1}{2}(y_2 + y_3) \right]$

Also, from the problem statement, we have

$$\overrightarrow{AP} = \frac{2}{3} \overrightarrow{AQ}$$

Hence, $(x - x_1) = \frac{2}{3}\left[\frac{1}{2}(x_2 + x_3) - x_1 \right]$

$$= \frac{2}{3}\left[\frac{x_2 + x_3 - 2x_1}{2} \right]$$

$$= \frac{1}{3}(x_2 + x_3 - 2x_1)$$

Therefore $x = \frac{1}{3}\left[(x_2 + x_3 - 2x_1) \right] + x_1$

$$= \frac{x_2 + x_3 - 2x_1 + 3x_1}{3}$$

$$= \frac{1}{3}(x_1 + x_2 + x_3)$$

Similarly, $(y - y_1) = \frac{2}{3}\left[\frac{1}{2}(y_2 + y_3) - y_1 \right]$

and $y = \frac{1}{3}(y_1 + y_2 + y_3)$

Therefore $(x, y) = \left\{ \left[\frac{1}{3}(x_1 + x_2 + x_3) \right], \left[\frac{1}{3}(y_1 + y_2 + y_3) \right] \right\}$ Ans.

We note from the above that even if a different vertex (say B or C) had been used, the result would have been the same due to the symmetry in the formulas for (x,y) representing the point P. Hence, we can say that the other two medians, with vertices at B and C, also pass through P.

VECTOR ADDITION
AND SCALAR MULTIPLICATION

■■ **1-3**

Show, using vectors, that the diagonals of a parallelogram bisect each other.

****

This diagonal is $\vec{u}+\vec{v}$; its midpoint is $\dfrac{\vec{u}+\vec{v}}{2}$.

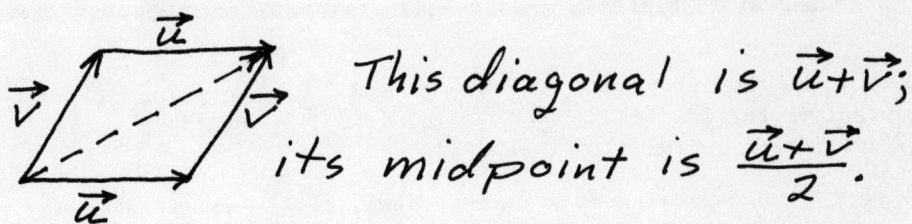

The midpoint of the other diagonal is given by $\vec{u}+\dfrac{\vec{v}-\vec{u}}{2} = \vec{u}+\dfrac{\vec{v}}{2}-\dfrac{\vec{u}}{2}$

$= \dfrac{\vec{u}}{2}+\dfrac{\vec{v}}{2} = \dfrac{\vec{u}+\vec{v}}{2}.$

1-4 ■■■

Given the points P(-2,5) and Q(3,-3), find the unit vector in the
direction of the displacement vector $\vec{PQ}$.

****

THE DISPLACEMENT VECTOR $\vec{PQ}$ IS
$$[3-(-2), -3-5] = [5,-8].$$
THE LENGTH (MAGNITUDE OF $\vec{PQ}$ IS
$$\sqrt{5^2 + (-8)^2} = \sqrt{89}.$$

THEREFORE, THE UNIT VECTOR IN THE DIRECTION
OF $\vec{PQ}$ IS
$$\frac{1}{\sqrt{89}}[5,-8] = \left[\frac{5}{\sqrt{89}}, \frac{-8}{\sqrt{89}}\right].$$

VECTOR FUNCTIONS

1-5 ■■■

Let C be the curve $x = t^3$, $y = t^4 + 3$. At the point (1, 4), the unit
tangent vector is (a) $3\vec{i} + 4\vec{j}$ (b) $1/\sqrt{3}\ \vec{i} + 4/(3\sqrt{3})\ \vec{j}$ (c) $3/5\ \vec{i} + 4/5\ \vec{j}$
$2/\sqrt{2}\ \vec{i} - 2/\sqrt{2}\ \vec{j}$ (e) $-4/5\ \vec{i} + 3/5\ \vec{j}$.

****

Let $\vec{R(t)} = \langle t^3, t^4+3 \rangle$, then $\vec{R'(t)} = \langle 3t^2, 4t^3 \rangle$

and so $\vec{R'(1)} = \langle 3,4 \rangle$. Now we need the unit vector

$\vec{T(1)}$ with the same direction as $\vec{R'(1)}$. $|\vec{R'(1)}| = \sqrt{9+16}$

$= \sqrt{25} = 5$, thus $\vec{T(1)} = \frac{1}{5}\vec{R'(1)} = \frac{1}{5}(3\vec{i}+4\vec{j}) = \frac{3}{5}\vec{i} + \frac{4}{5}\vec{j}.$

━━━━━━━━━━━━━━━━━━━━━━━━━━━━━━━━━━━━ **1-6**

Let $R(t) = 2t^2 i + \dfrac{1}{3t^3} j$. Find:

a. the corresponding parametric equations

b. $R'(t)$

c. $\dfrac{dy}{dx}$ (in terms of t)

d. $\dfrac{d^2 y}{dx^2}$ (in terms of t)

****

a. $x = 2t^2$ and $y = \dfrac{1}{3t^3}$ $\left(\text{since } R(t) = xi + yj \right)$

b. $R'(t) = 4ti - \dfrac{1}{t^4} j$

c. $\dfrac{dy}{dx} = \dfrac{\frac{dy}{dt}}{\frac{dx}{dt}} = \dfrac{-\frac{1}{t^4}}{4t} = -\dfrac{1}{4t^5}$

d. $\dfrac{d^2 y}{dx^2} = \dfrac{\frac{d\left(\frac{dy}{dx}\right)}{dt}}{\frac{dx}{dt}} = \dfrac{\frac{d\left(-\frac{1}{4t^5}\right)}{dt}}{\frac{dx}{dt}} = \dfrac{\frac{5}{4t^6}}{4t} = \dfrac{5}{16t^7}$

━━━━━━━━━━━━━━━━━━━━━━━━━━━━━━━━━━━━ **1-7**

Find the unit tangent and the unit normal to the graph of the vector valued function $\vec{r}(t) = \langle t^2 - 2,\ 2t - t^3 \rangle$ at t=1.

****

$r'(t) = \langle 2t,\ 2 - 3t^2 \rangle, \quad r'(1) = \langle 2, -1 \rangle.$

$\| r'(1) \| = \sqrt{5}$

$\therefore$ tangent $= \dfrac{1}{\sqrt{5}} \langle 2, -1 \rangle,$ normal $= \dfrac{1}{\sqrt{5}} \langle 1, 2 \rangle$

1-8

Find [f(t) . g(t)]' and [f(t) x g(t)]' for

$$f(t) = t^3 i - t^2 j + t k$$

and $$g(t) = t i - t^2 j + t^3 k$$

****

From the given equations for $f(t)$ and $g(t)$, we write:

$$[f(t) \cdot g(t)] = (t^3 \cdot t)\hat{i} + [(-t^2) \cdot (-t^2)]\hat{j} + (t \cdot t^3)\hat{k}$$

$$= t^4 \hat{i} + t^4 \hat{j} + t^4 \hat{k}$$

$$= t^4 + t^4 + t^4 = 3t^4$$

Hence, $\left[f(t) \cdot g(t)\right]' = 12t^3$ <u>Ans.</u>

Similarly, for $\left[f(t) \times g(t)\right]$ we write:

$$f'(t) = 3t^2\hat{i} - 2t\hat{j} + 1\hat{k}$$

and $$g'(t) = 1\hat{i} - 2t\hat{j} + 3t^2\hat{k}$$

Now, $\left[f(t) \times g(t)\right]' = \left[f'(t) \times g(t)\right] + \left[f(t) \times g'(t)\right]$

$$= \begin{vmatrix} \hat{i} & \hat{j} & \hat{k} \\ 3t^2 & -2t & 1 \\ t & -t^2 & t^3 \end{vmatrix} + \begin{vmatrix} \hat{i} & \hat{j} & \hat{k} \\ t^3 & -t^2 & t \\ 1 & -2t & 3t^2 \end{vmatrix}$$

$$= \left[(-2t^4 + t^2)i - (3t^5 - t)j + (-3t^4 + 2t^2)k\right]$$

$$+ \left[(-3t^4 + 2t^2)i - (3t^5 - t)j + (-2t^4 + t^2)k\right]$$

$$= (-5t^4 + 3t^2)i - (6t^5 - 2t)j + (-5t^4 + 3t^2)k \quad \underline{Ans.}$$

MOTION IN THE PLANE

━━ **1-9**

Suppose a particle moves in the plane according to the vector-valued function $f(t) = 2e^t \vec{i} + e^{-t} \vec{j}$, where t represents time. Find $v(t)$, $\left| v(t) \right|$, $a(t)$, and sketch a graph showing the path taken by the particle indicating the direction of the motion.

****

$$f(t) = 2e^t \vec{i} + e^{-t} \vec{j}$$

$$v(t) = f'(t) = 2e^t \vec{i} - e^{-t} \vec{j}$$

$$\left| v(t) \right| = \sqrt{(2e^t)^2 + (-e^{-t})^2} = \sqrt{4e^{2t} + e^{-2t}}$$

$$a(t) = v'(t) = 2e^t \vec{i} + e^{-t} \vec{j}$$

to sketch, break into parametric form.

$$x = 2e^t \qquad y = e^{-t} = 1/e^t = 1/x_{/2} = 2/x \Rightarrow xy = 2$$
$$x_{/2} = e^t$$

as $t \to +\infty$, $x \to +\infty$ and $y \to 0$ so direction is from left to right along Q I branch of hyperbola.

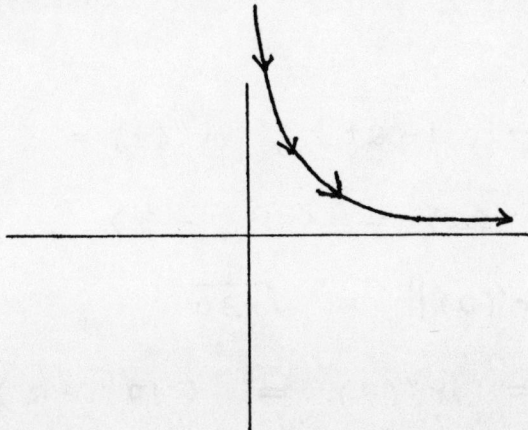

1-10 ■■■

An object's postion vector at time t is given by the vector-valued function $\mathbf{F}(t) = [t^3-t, 2t^2+t]$. At t=2, find its velocity vector and its speed.

****

THE VELOCITY VECTOR IS GIVEN BY
$$V(t) = F'(t) = [3t^2-1, \ 4t+1].$$
AT $t=2$,

THE VELOCITY VECTOR IS
$$V(2) = [3\cdot2^2-1, \ 4\cdot2+1] = [11, 9]$$
AND THE SPEED IS
$$|V(2)| = \sqrt{11^2+9^2} = \sqrt{202}.$$

1-11 ■■■

Suppose a particle is moving in the xy-plane so that its position vector at time t is given by $\vec{r}(t) = \langle t^3-t, \ t-t^2 \rangle$. Find the velocity, speed, and acceleration of the particle at time t=2.

****

$$r'(t) = \langle 3t^2-1, \ 1-2t \rangle, \quad r''(t) = \langle 6t, -2 \rangle$$

$$velocity = r'(2) = \langle 11, -3 \rangle$$

$$speed = \|r'(2)\| = \sqrt{130}$$

$$acceleration = r''(2) = \langle 12, -2 \rangle$$

━━━ **1-12**

The position of a particle at time t is given parametrically by

$$y = t^2 \quad \text{and} \quad x = \frac{1}{3}(t^3 - 3t).$$

a) Show that the particle crosses the y-axis three times.

b) Find dy/dx and use the answer to find where the path has horizontal and vertical tangents.

****

a) The particle crosses the y-axis when $x = 0$

$$x = 0 \Rightarrow 0 = \tfrac{1}{3}(t^3 - 3t) \Rightarrow 0 = t(t^2 - 3)$$

$$\Rightarrow t = -\sqrt{3}, \; 0, \; \sqrt{3}$$

Hence there are three crossings.

b) Now $\dfrac{dy}{dt} = 2t$ and $\dfrac{dx}{dt} = t^2 - 1$

So $\dfrac{dy}{dx} = \dfrac{dy/dt}{dx/dt} = \dfrac{2t}{t^2 - 1}$

Horizontal tangents occur when $\dfrac{dy}{dx} = 0$.

$$\dfrac{dy}{dx} = 0 \Rightarrow 2t = 0 \Rightarrow \underline{t = 0.}$$

Vertical tangents occur when $\dfrac{dy}{dx}$ is undefined.

$$\dfrac{dy}{dx} \text{ undefined} \Rightarrow t^2 - 1 = 0 \Rightarrow \underline{\underline{t = \pm 1.}}$$

1-13 ■■■

Let $\vec{A(t)}$, $\vec{V(t)}$, $\vec{R(t)}$ denote respectively the acceleration, velocity, and position at time t of an object moving in the xy plane.
Find $\vec{R(t)}$, given $\vec{A(t)} = \langle e^{2t} + 2t, e^{2t} - 3\rangle$, $\vec{V(0)} = \langle 3/2, 7/2\rangle$, and $\vec{R(0)} = \langle 5/4, 9/4\rangle$.

****

$$\vec{V(t)} = \int \vec{A(t)}\, dt = \langle \tfrac{1}{2}e^{2t} + t^2 + C_1, \tfrac{1}{2}e^{2t} - 3t + C_2\rangle. \text{ To}$$

determine C_1, C_2, let $t=0$, $\vec{V(0)} = \langle \tfrac{1}{2} + C_1, \tfrac{1}{2} + C_2\rangle = \langle \tfrac{3}{2}, \tfrac{7}{2}\rangle$,

$\tfrac{1}{2} + C_1 = \tfrac{3}{2}$, $C_1 = 1$; $\tfrac{1}{2} + C_2 = \tfrac{7}{2}$, $C_2 = 3$, thus

$$\vec{V(t)} = \langle \tfrac{1}{2}e^{2t} + t^2 + 1, \tfrac{1}{2}e^{2t} - 3t + 3\rangle.$$

$$\vec{R(t)} = \int \vec{V(t)}\, dt = \langle \tfrac{1}{4}e^{2t} + \tfrac{1}{3}t^3 + t + C_3, \tfrac{1}{4}e^{2t} - \tfrac{3}{2}t^2 + 3t + C_4\rangle,$$

$\vec{R(0)} = \langle \tfrac{1}{4} + C_3, \tfrac{1}{4} + C_4\rangle = \langle \tfrac{5}{4}, \tfrac{9}{4}\rangle$, $\tfrac{1}{4} + C_3 = \tfrac{5}{4}$, $C_3 = 1$;

$\tfrac{1}{4} + C_4 = \tfrac{9}{4}$, $C_4 = 2$. Therefore,

$$\vec{R(t)} = \langle \tfrac{1}{4}e^{2t} + \tfrac{1}{3}t^3 + t + 1, \tfrac{1}{4}e^{2t} - \tfrac{3}{2}t^2 + 3t + 2\rangle.$$

━━━━━━━━━━━━━━━━━━━━━━━━━━━**1-14**

A paper boy is traveling 60 miles per hour down a straight road in the
direction of the vector i when he throws a paper out the car window with
a velocity relative to the car, in the direction of j and of magnitude 10
miles per hour. (a) Find the speed of the paper relative to the ground
when the boy lets it go. (b) Find the velocity of the paper at that
time.

****

The information above may be illustrated
by the following diagram:

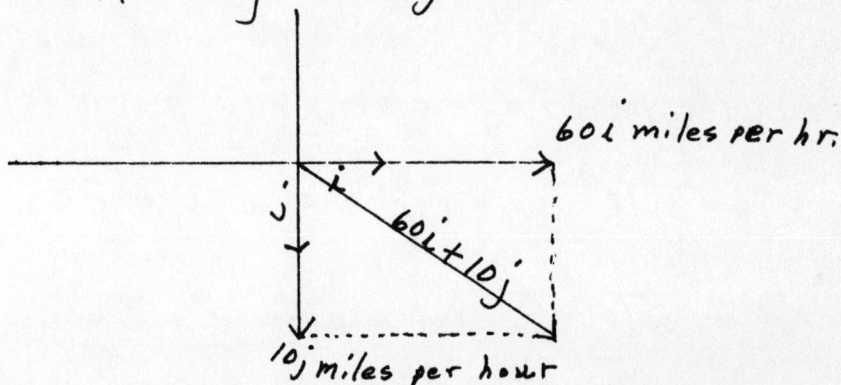

$60i$ miles per hr.

$60i + 10j$

$10j$ miles per hour

(a) Let V denote the velocity relative to
the ground. The V is the resultant
of the vectors $60i$ and $10j$; hence,

$$V = 60i + 10j$$

(b) By definition the speed is the
length of V. So Speed $= \sqrt{(60)^2 + (10)^2}$

$$= \sqrt{3700} \text{ miles per hour}$$

1-15

A particle has velocity given by $\vec{v}(t) = e^t\,\vec{i} + 2\vec{j}$. At time $t = 0$ it is at the origin. Find each of the following:
(a) its speed and (b) its distance from the origin at time $t = 1$.

****

a) speed $= \|\vec{v}(t)\| = \|e^t\vec{i} + 2\vec{j}\| = \sqrt{e^{2t} + 4}$

at time $t = 1$, its speed $= \sqrt{e^2 + 4}$

b) the position vector $\vec{r}(t) = \int \vec{v}(t)\,dt$

$$= \int (e^t\vec{i} + 2\vec{j})\,dt = e^t\vec{i} + 2t\vec{j} + \vec{c}$$

Since $\vec{r}(0) = \vec{0}$, we have $\vec{i} + \vec{c} = \vec{0}$, or $\vec{c} = -\vec{i}$

Hence $\vec{r}(t) = (e^t\vec{i} + 2t\vec{j}) - \vec{i}$

$$= (e^t - 1)\vec{i} + 2t\vec{j}$$

and $\vec{r}(1) = (e-1)\vec{i} + 2\vec{j}$

$\therefore$ distance from $(0,0)$ is given by:

$$\|\vec{r}(1)\| = \sqrt{(e-1)^2 + 4} = \sqrt{e^2 - 2e + 5}$$

■■■ **1-16**

A cannon sits on top of a vertical tower 264 feet tall. It fires a
cannonball at 80 ft/sec. If the barrel of the cannon is elevated 30
degrees from horizontal, find how far from the base of the tower the
cannonball will land (assuming the ground around the tower is level).

****

Acceleration vector $\vec{A}(t) = -32\vec{j}$ (ft/sec²)

Velocity vector $\vec{V}(t) = c_1\vec{i} + (-32t + k_1)\vec{j}$ (ft/sec)

Position vector $\vec{R}(t) = (c_1 t + c_2)\vec{i} + (-16t^2 + k_1 t + k_2)\vec{j}$ (feet)

Initial Velocity   Initial Position

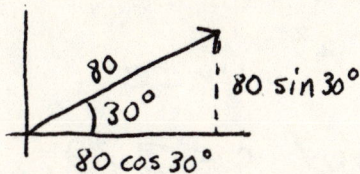

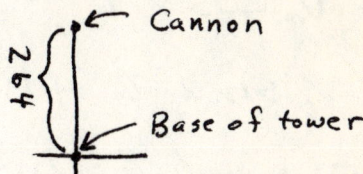

$\vec{V}(0) = 40\sqrt{3}\,\vec{i} + 40\vec{j}$ $\vec{R}(0) = 0\vec{i} + 264\vec{j}$

∴ $c_1 = 40\sqrt{3}$, $k_1 = 40$ ∴ $c_2 = 0$, $k_2 = 264$

Thus, $\vec{R}(t) = (40\sqrt{3}\,t)\vec{i} + (-16t^2 + 40t + 264)\vec{j}$

Finding the value of t when $-16t^2 + 40t + 264 = 0$ gives
the time when the cannonball lands.

$-16t^2 + 40t + 264 = -8(2t^2 - 5t - 33) = -8(2t-11)(t+3) = 0$

∴ $t = -3$ or $t = \frac{11}{2}$. Since we start at time $t = 0$ and

consider increasing time, we don't consider $t = -3$. The
desired time is $t = \frac{11}{2}$

The $\vec{i}$ component of $\vec{R}(t)$, $40\sqrt{3}\,t$, measures the
horizontal distance travelled at time t.

∴ $40\sqrt{3}\left(\frac{11}{2}\right) = 220\sqrt{3}$ ft is the distance from

the base of the tower.

1-17 ■■■

A person is standing 80 feet from a tall cliff. She throws a rock at 80 feet per second at an angle of 45°. Neglecting air resistance, how far up the cliff does it hit?

****

If the person's height is h, the equations of the trajectory are

$$x(t) = V_0 \cos\alpha\, t,$$

$$y(t) = V_0 \sin\alpha\, t + h - \frac{1}{2}gt^2.$$ Since $V_0 = 80$ and $\alpha = 45°$ (and $g = 32$),

$$x = \frac{80t}{\sqrt{2}}, \quad y = \frac{80t}{\sqrt{2}} + h - 16t^2.$$

The rock hits the cliff when $x = 80$, i.e. $t = \sqrt{2}$ seconds. When $t = \sqrt{2}$,

$$y = 80 + h - 32 = 48 + h \text{ feet up the cliff.}$$

■■■ **1-18**

If $R(t) = (t^2 + 3)i + (2t^2 - 3t + 5)j$ describes the motion of a particle, find:

a. the velocity when t = 3.
b. the speed when t = 3.
c. the acceleration when t = 3.

****

a. Since $V(t) = R'(t)$, $V(t) = 2ti + (4t-3)j$.
   So $V(3) = 6i + 9j$

b. Since the speed is the norm (or magnitude) of the velocity,
$$\|V(3)\| = \|6i + 9j\| = \sqrt{6^2 + 9^2} = \sqrt{117} = 3\sqrt{13}$$

c. $A(t) = V'(t) = R''(t)$, so in this case
   $A(t) = 2i + 4j$ (note that this is constant).
   So $A(3) = 2i + 4j$

LENGTH OF ARC

1-19 ▰▰▰▰▰▰▰▰▰▰▰▰▰▰▰▰▰▰▰▰▰▰▰▰▰▰▰▰▰▰▰

Find the length of the curve r = $\sin^3\left(\frac{\theta}{3}\right)$.

****

$$L = \int_0^{2\pi} \sqrt{\left(\sin^3\left(\frac{t}{3}\right)\right)^2 + \left(3\sin^2\left(\frac{t}{3}\right)\cos\left(\frac{t}{3}\right)\frac{1}{3}\right)^2}\, dt =$$

$$\int_0^{2\pi} \sqrt{\sin^6\left(\frac{t}{3}\right) + \sin^4\left(\frac{t}{3}\right)\cos^2\left(\frac{t}{3}\right)}\, dt =$$

$$\int_0^{2\pi} \sqrt{\sin^4\left(\frac{t}{3}\right)}\sqrt{\sin^2\left(\frac{t}{3}\right) + \cos^2\left(\frac{t}{3}\right)}\, dt = \int_0^{2\pi}\sin^2\left(\frac{t}{3}\right)\, dt =$$

$$\frac{1}{2}\int_0^{2\pi} 1 - \cos\left(2\frac{t}{3}\right)\, dt = \frac{1}{2}\left\{ t - \frac{3\sin\frac{2t}{3}}{2}\Big|_0^{2\pi}\right\} =$$

$$\frac{1}{2}\left\{\left(2\pi - \frac{3}{2}\sin\frac{4\pi}{3}\right) - \left(0 - \frac{3}{2}\sin 0\right)\right\} =$$

$$\frac{1}{2}\left\{2\pi + \frac{3}{2}\frac{\sqrt{3}}{2}\right\} = \pi + \frac{3\sqrt{3}}{8}.$$

■■1-20

Find the length of the curve represented by:

$$x = \text{Arcsin}(t/2) \quad \text{and} \quad y = \ln\sqrt{4 - t^2} \qquad 0 \leq t \leq 1$$

****

$$L = \int_a^b \sqrt{[x'(t)]^2 + [y'(t)]^2}\ dt$$

$$x'(t) = \frac{1}{\sqrt{1 - (t/2)^2}} \cdot \frac{1}{2} = \frac{1}{\sqrt{4 - t^2}}$$

$$y'(t) = \frac{(1/2)(4 - t^2)^{-1/2}(-2t)}{(4 - t^2)^{1/2}} = \frac{-t}{4 - t^2}$$

$$\text{so } L = \int_0^1 \left(\left[\frac{1}{\sqrt{4 - t^2}}\right]^2 + \left[\frac{-t}{4 - t^2}\right]^2 \right)^{1/2} dt$$

$$= \int_0^1 \left(\frac{1}{4 - t^2} + \frac{t^2}{(4 - t^2)^2} \right)^{1/2} dt$$

$$= \int_0^1 \left(\frac{(4 - t^2) + t^2}{(4 - t^2)^2} \right)^{1/2} dt = \int_0^1 \frac{2}{4 - t^2}\ dt$$

by partial fractions $= \int_0^1 \frac{\frac{1}{2}}{2 + t} + \frac{\frac{1}{2}}{2 - t}\ dt$

$$= \frac{1}{2}\ln|2 + t| - \frac{1}{2}\ln|2 - t|\ \Big]_0^1$$

$$= \ln\left(\frac{2 + t}{2 - t}\right)^{1/2} = \ln\sqrt{3} - \ln 1 = \ln\sqrt{3}$$

1-21 ■■■

Find the circumference of the circle: x = 2 cos t, y = 2 sin t

****

$$\frac{dx}{dt} = -2\sin t \quad ; \quad \frac{dy}{dt} = 2\cos t$$

Since arc length is defined as

$$S = \int_{t_1}^{t_2} \sqrt{\left(\frac{dx}{dt}\right)^2 + \left(\frac{dy}{dt}\right)^2} \, dt \quad ,$$

$$S = \int_0^{2\pi} \sqrt{(-2\sin t)^2 + (2\cos t)^2} \, dt$$

$$= \int_0^{2\pi} \sqrt{4\sin^2 t + 4\cos^2 t} \, dt$$

$$= \int_0^{2\pi} \sqrt{4(\sin^2 t + \cos^2 t)} \, dt$$

$$= \int_0^{2\pi} \sqrt{4} \, dt$$

$$= \int_0^{2\pi} 2 \, dt$$

$$= 2t \Big|_0^{2\pi}$$

$$= 4\pi$$

1-22

If x = cos(2t), y = sin^{2}t and (x,y) represents the position of a particle, find the distance the particle travels as t moves from 0 to $\frac{\pi}{2}$.

****

Let $x = f(t) = \cos 2t$ and $y = g(t) = \sin^2 t$ and we will use $L = \int_0^{\frac{\pi}{2}} \sqrt{f'(t)^2 + g'(t)^2}\, dt$ to find the distance. Now $f'(t) = -2 \sin 2t$ and $g'(t) = 2 \sin t \cos t = \sin 2t$.

$$L = \int_0^{\frac{\pi}{2}} \sqrt{4\sin^2 2t + \sin^2 2t}\, dt = \sqrt{5} \int_0^{\frac{\pi}{2}} |\sin 2t|\, dt.$$

Since $\sin 2t \geq 0$ for all $t \in [0, \frac{\pi}{2}]$ we have $L = \sqrt{5} \int_0^{\frac{\pi}{2}} \sin 2t\, dt = \frac{\sqrt{5}}{2} \int_0^{\frac{\pi}{2}} \sin 2t\, (2\,dt) =$

$$\frac{\sqrt{5}}{2} \left\{ -\cos 2t \Big|_0^{\frac{\pi}{2}} \right\} = \frac{\sqrt{5}}{2} \left\{ -\cos \pi + \cos 0 \right\} =$$

$$\frac{\sqrt{5}}{2} \left\{ -(-1) + 1 \right\} = \frac{2\sqrt{5}}{2} = \sqrt{5}.$$

1-23 ■■■

Consider the plane curve given by the parameters

$$x(t) = t^3$$
$$y(t) = t^2$$

Find the length of the curve from the point where t = 0 to the point where t = 1.

****

Letting s denote the length of the curve :

$$s = \int_0^1 \sqrt{[x'(t)]^2 + [y'(t)]^2} \, dt$$

$$= \int_0^1 \sqrt{9t^4 + 16t^2} \, dt$$

$$= \int_0^1 t\sqrt{9t^2 + 16} \, dt$$

$$= \frac{1}{27} (9t^2 + 16)^{3/2} \Big|_0^1$$

$$= \frac{125}{27} - \frac{64}{27} = \boxed{\frac{61}{27}}$$

■■■ 1-24

The involute of a circle of radius 1 is given parametrically by:

$$x = \cos t + t \sin t$$
$$y = \sin t - t \cos t$$

Find the length of the portion of the involute which is traced out as t increases from 0 to π.

****

Recall the arc length formula $\quad L = \int_a^b \sqrt{\left(\dfrac{dx}{dt}\right)^2 + \left(\dfrac{dy}{dt}\right)^2}\ dt$

Now $\quad \dfrac{dx}{dt} = -\sin t + t\cos t + \sin t = t\cos t$

and $\quad \dfrac{dy}{dt} = \cos t - (-t\sin t + \cos t) = t\sin t$

So $\quad L = \int_0^\pi \sqrt{(t\cos t)^2 + (t\sin t)^2}\ dt$

$$= \int_0^\pi t\sqrt{\cos^2 t + \sin^2 t}\ dt$$

$$= \int_0^\pi t\ dt \quad = \underline{\underline{\dfrac{\pi^2}{2}}}$$

1-25 ▬▬▬▬▬▬▬▬▬▬▬▬▬▬▬▬▬▬▬▬▬▬▬▬▬

Given the parametric equations

$$x = 2 + 2 \sin t$$
$$y = 5 - 2 \cos t$$

rewrite them as parametric equations where s (the arc length) is the parameter. s = 0 when t = 0.

****

Since $\dfrac{ds}{dt} = \sqrt{\left(\dfrac{dx}{dt}\right)^2 + \left(\dfrac{dy}{dt}\right)^2}$ we find

$$\dfrac{dx}{dt} = 2 \cos t \quad \text{and} \quad \dfrac{dy}{dt} = 2 \sin t$$

Thus $\dfrac{ds}{dt} = \sqrt{(2\cos t)^2 + (2 \sin t)^2} = \sqrt{4\cos^2 t + 4 \sin^2 t}$

$$= \sqrt{4(\cos^2 t + \sin^2 t)} = \sqrt{4 \cdot 1} = 2$$

Since $\dfrac{ds}{dt} = 2$, we integrate to get $s = 2t + c$

Because $s = 0$ when $t = 0$, $0 = 2(0) + c$ so $c = 0$

and $s = 2t$, so $t = \dfrac{s}{2}$.

Therefore the answer is

$$x = 2 + 2 \sin \dfrac{s}{2}$$
$$y = 5 - 2 \cos \dfrac{s}{2}$$

■■■**1-26**

Compute the length of the curve given parametrically by $x = (1/3)t^3$, $y = (1/2)t^2$ for $0 \leq t \leq 2$.

****

$$\frac{dx}{dt} = t^2 \qquad \frac{dy}{dt} = t$$

$$\int_0^2 \sqrt{(t^2)^2 + (t)^2}\, dt = \int_0^2 (t^4 + t^2)^{\frac{1}{2}}\, dt$$

$$= \int_0^2 (t^2 + 1)^{\frac{1}{2}} t\, dt \quad \text{SINCE } t \geq 0$$

$$= \frac{1}{2} \int_0^2 (t^2 + 1)^{\frac{1}{2}} 2t\, dt$$

$$= \frac{1}{2} \left. \frac{(t^2 + 1)^{\frac{3}{2}}}{\frac{3}{2}} \right|_0^2$$

$$= \frac{1}{3} \left. (t^2 + 1)^{\frac{3}{2}} \right|_0^2$$

$$= \frac{1}{3}(2^2 + 1)^{\frac{3}{2}} - \frac{1}{3}(0^2 + 1)^{\frac{3}{2}}$$

$$= \frac{5\sqrt{5} - 1}{3}$$

■■■**1-27**

Find the length of the curve $x = t^2/2 + 7$, $y = 1/3\,(2t+1)^{3/2}$, $2 \leq t \leq 6$.

****

$$x'(t) = t, \quad y'(t) = \frac{1}{3} \cdot \frac{3}{2}(2t+1)^{\frac{1}{2}} \cdot 2 = \sqrt{2t+1}. \text{ Therefore,}$$

$$L = \int_2^6 \sqrt{x'(t)^2 + y'(t)^2}\, dt = \int_2^6 \sqrt{t^2 + 2t + 1}\, dt = \int_2^6 \sqrt{(t+1)^2}\, dt$$

$$= \int_2^6 (t+1)\, dt = \left. \frac{t^2}{2} + t \right|_2^6 = 18 + 6 - (2 + 2) = 20.$$

1-28 ▪▪▪▪▪▪▪▪▪▪▪▪▪▪▪▪▪▪▪▪▪▪▪▪▪▪▪▪▪▪▪▪▪▪▪▪▪▪

An arc is described by the parametric equations $x = 3t^3$ and $y = 2t^2$. Sketch the arc and calculate its length from (a) t = 0 to t = 2, and (b) t = -3 to t = 0.

****

By choosing selected values for t we obtain the following coordinates, which are used to draw the graph shown alongside.

t	0	1	2	2.5	3	-1	-2	-3
x	0	3	24	46.8	81	-3	-24	-81
Y	0	2	8	12.5	18	2	8	18

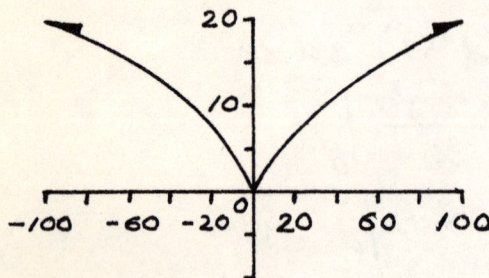

Now, Let:

$f(t) = x = 3t^3$, and

$g(t) = y = 2t^2$

Hence,

$f'(t) = 9t^2$; $g'(t) = 4t$

For calculating the length of arc from t 0 to t = 2 we write

$$L = \int_a^b \sqrt{[f'(t)]^2 + [g'(t)]^2}\, dt$$

$$= \int_0^2 \sqrt{(9t^2)^2 + (4t)^2}\, dt$$

$$= \int_0^2 \sqrt{81t^4 + 16t^2}\, dt = \int_0^2 t\sqrt{81t^2 + 16}\, dt$$

Now, let $u = 81t^2 + 16$

$\frac{du}{dt} = 81(2)t$; $du = 162t\, dt$; $\frac{du}{162} = t \cdot dt$

Hence, $L = \int_0^2 \frac{1}{162} u^{1/2}\, du$

$$= \frac{1}{162} \cdot \frac{2}{3} (81t^2 + 16)^{3/2} \Big/_0^2$$

$$= \frac{1}{243} \left[\{81(2)^2 + 16\}^{3/2} - \{81(0)^2 + 16\}^{3/2} \right]$$

$$= \frac{1}{243} \left[340^{3/2} - 16^{3/2} \right] = 25.53 \text{ units} \quad \underline{\text{Ans.}}$$

For computing the length of arc from t = -3 to t = 0, we similarly write:

$$L = \int_{-3}^{0} \sqrt{81 t^4 + 16 t^2} \; dt$$

$$= \int_{-3}^{0} \sqrt{t^2} \; \sqrt{81 t^2 + 16} \; dt$$

Now, for $-3 \leq t \leq 0$, $\sqrt{t^2} = -t$. Hence,

$$L = \int_{-3}^{0} -t \sqrt{81 t^2 + 16} \; dt$$

$$= -\frac{1}{243} \left(81 t^2 + 16 \right)^{3/2} \Big|_{-3}^{0}$$

$$= -\frac{1}{243} \left[\left\{ 81 (0)^2 + 16 \right\}^{3/2} - \left\{ 81 (-3)^2 + 16 \right\}^{3/2} \right]$$

$$= -\frac{1}{243} \left[16^{3/2} - 745^{3/2} \right] = 83.41 \text{ units}$$
$$\underline{Ans.}$$

■■1-29

Find the arc length of the curve r = 3sinθ.

****

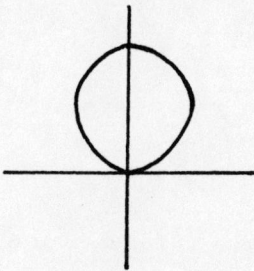

$$S = \int_{a}^{b} \sqrt{r^2 + \left(\frac{dr}{d\theta}\right)^2} \; d\theta = \int_{0}^{\pi} \sqrt{9\sin^2\theta + 9\cos^2\theta} \; d\theta$$

$$= \int_{0}^{\pi} 3 \, d\theta = 3\theta \Big|_{0}^{\pi} = \underline{3\pi \text{ units}}$$

1-30 ██

The equation of a curve in parametric form is

$$x = 4 \cos 3t, \quad y = 4 \sin 3t.$$

1. Find a corresponding Cartesian equation of the curve.
2. Find the arc length of the curve from $t = 0$ to $t = \pi/8$.

****

1. Eliminate t by squaring both equations and adding.

$$x = 4\cos 3t \Rightarrow x^2 = 16 \cos^2 3t$$
$$y = 4\sin 3t \Rightarrow y^2 = 16 \sin^2 3t$$
$$x^2 + y^2 = 16(\cos^2 3t + \sin^2 3t)$$
$$x^2 + y^2 = 16$$

2. $\frac{dx}{dt} = 4(-\sin 3t)(3) = -12\sin 3t, \quad \frac{dy}{dt} = 4(\cos 3t)(3)$
$= 12\cos 3t$. Then

$$\text{Arc length} = \int_a^b \sqrt{\left(\frac{dx}{dt}\right)^2 + \left(\frac{dy}{dt}\right)^2} \, dt$$

$$= \int_0^{\pi/8} \sqrt{(-12\sin 3t)^2 + (12\cos 3t)^2} \, dt = \int_0^{\pi/8} \sqrt{144(\sin^2 3t + \cos^2 3t)} \, dt$$

$$= \int_0^{\pi/8} \sqrt{144} \, dt = 12t \Big|_0^{\pi/8} = 12\left(\frac{\pi}{8}\right) = \frac{3\pi}{2}$$

■■ **1-31**

Find the length of the plane curve x(t) = t, y(t) = $t^{2/3}$ between t = 0 and
t = 8.

****

$$Length = \int_{t_0}^{t_1} \sqrt{\left(\frac{dx}{dt}\right)^2 + \left(\frac{dy}{dy}\right)^2}\, dt$$

$$= \int_0^8 \sqrt{1 + \left(\frac{2}{3}t^{-1/3}\right)^2}\, dt$$

$$= \int_0^8 \sqrt{1 + \frac{4}{9t^{2/3}}}\, dt = \int_0^8 \frac{1}{3t^{1/3}}\sqrt{9t^{2/3}+4}\, dt$$

$$= \left(\text{if we let } u = 4 + 9t^{2/3}, du = 6t^{-\frac{1}{3}}dt\right)$$

$$\int_4^{40} \frac{1}{18}\sqrt{u}\, du = \frac{2}{3}\cdot\frac{1}{18}\cdot u^{3/2}\Big|_4^{40}$$

$$= \frac{1}{27}\left(40^{3/2} - 4^{3/2}\right)$$

$$= \frac{8}{27}\left(10\sqrt{10} - 1\right).$$

1-32 ██

A curve is written parametrically as $x = t^3$, $y = t^{9/2}$. Find the length of the curve from t=0 to t=1.

****

Arc length $s = \int_{t_1}^{t_2} \sqrt{\left(\frac{dx}{dt}\right)^2 + \left(\frac{dy}{dt}\right)^2}\, dt$ $\frac{dx}{dt} = 3t^2$, $\frac{dy}{dt} = \frac{9}{2}t^{7/2}$

$\therefore s = \int_0^1 \sqrt{(3t^2)^2 + \left(\frac{9}{2}t^{7/2}\right)^2}\, dt = \int_0^1 \sqrt{9t^4 + \frac{81}{4}t^7}\, dt$

$= \int_0^1 \sqrt{9t^4\left(1 + \frac{9}{4}t^3\right)}\, dt = \int_0^1 3t^2 \sqrt{1 + \frac{9}{4}t^3}\, dt$

$\left(\text{let } u = 1 + \frac{9}{4}t^3, \quad du = \frac{27}{4}t^2\, dt, \quad t^2 dt = \frac{4}{27}\, du \right)$

$= \int_{u_1}^{u_2} 3\sqrt{u}\, \frac{4}{27}\, du = \frac{4}{9}\, \frac{u^{3/2}}{3/2}\bigg|_{u_1}^{u_2} = \frac{8}{27}\left[\left(1 + \frac{9}{4}t^3\right)^{3/2}\right]_0^1$

$= \frac{8}{27}\left[\left(1 + \frac{9}{4}(1)^3\right)^{3/2} - \left(1 + \frac{9}{4}(0)^3\right)^{3/2}\right] = \frac{8}{27}\left[\left(\frac{13}{4}\right)^{3/2} - 1\right]$

$= \frac{8}{27}\left[\frac{13}{8}\sqrt{13} - 1\right]$

━━━━━━━━━━━━━━━━━━━━━━━━━━━━━━━━ **1-33**

Find the arc length of the curve defined by:

$$x = t^3 + 1, \ y = 3t^2 + 2, \ \text{for } 0 \le t \le 1.$$

****

$$\text{Arc length} = \int_0^1 \sqrt{\left(\frac{dx}{dt}\right)^2 + \left(\frac{dy}{dt}\right)^2} \ dt$$

$$= \int_0^1 \sqrt{(3t^2)^2 + (6t)^2} \ dt = \int_0^1 \sqrt{9t^4 + 36t^2} \ dt$$

$$= 3 \int_0^1 t\sqrt{t^2 + 4} \ dt = \left. (t^2 + 4)^{3/2} \right|_0^1$$

$$= 5\sqrt{5} - 8$$

TANGENTIAL AND NORMAL COMPONENTS
OF ACCELERATION

1-34 ■■

A projectile moves according to the vector function $\vec{R}(t) = t^2 \vec{i} + (3t-2) \vec{j}$. Find the tangential and normal components of acceleration for this projectile at the time t=2.

****

Normal component $A_N(t) = \left(\frac{ds}{dt}\right)^2 k$, where k is the curvature and $\frac{ds}{dt}$ is the speed.

$$\frac{ds}{dt} = |V(t)| = |R'(t)| = \sqrt{(2t)^2 + (3)^2} = \sqrt{4t^2 + 9}$$

$$k = \frac{\frac{dx}{dt} \cdot \frac{d^2y}{dt^2} - \frac{dy}{dt} \cdot \frac{d^2x}{dt^2}}{\left[\left(\frac{dx}{dt}\right)^2 + \left(\frac{dy}{dt}\right)^2\right]^{3/2}} = \frac{2t(0) - 3(2)}{\left[(2t)^2 + (3)^2\right]^{3/2}} = \frac{-6}{(4t^2+9)^{3/2}}$$

$$A_N(t) = \frac{\left(\sqrt{4t^2+9}\right)^2 (-6)}{(4t^2+9)^{3/2}} = \frac{-6}{\sqrt{4t^2+9}}, \quad A_N(2) = -\frac{6}{5}$$

Tangential component $A_T(t) = \frac{d^2s}{dt^2} = \frac{d}{dt}\left(\sqrt{4t^2+9}\right)$

$$= \frac{1}{2}(4t^2+9)^{-1/2}(8t) = \frac{4t}{\sqrt{4t^2+9}}, \quad A_T(2) = \frac{8}{5}$$

--**1-35**

The position of a particle at time t is given by

$$\vec{r}(t) = e^t\vec{\imath} + t\vec{\jmath}.$$

Find the tangential and normal components of acceleration.

****

To find the tangential component of acceleration (a_T), recall that $a_T = d^2s/dt^2$.

But $ds/dt = \left\| d\vec{r}/dt \right\| = \left\| e^t\vec{\imath} + \vec{\jmath} \right\| = (e^{2t} + 1)^{1/2}$

So $a_T = \dfrac{d}{dt}\left[(e^{2t} + 1)^{1/2} \right] = \dfrac{e^{2t}}{(e^{2t} + 1)^{1/2}}$

For the normal component of acceleration (a_N), recall that $a_N = \left(\|\vec{a}\|^2 - (a_T)^2 \right)^{1/2}$

But $\|\vec{a}\| = \left\| d^2\vec{r}/dt^2 \right\| = \left\| e^t\vec{\imath} \right\| = e^t$

So $a_N = \left(e^{2t} - \dfrac{e^{4t}}{e^{2t} + 1} \right)^{1/2} = \dfrac{e^{2t}}{(e^{2t} + 1)^{1/2}}$

Note that for this problem $a_T = a_N$.

1-36 ■■■

The position vector of a particle at time t is $\underline{r}(t) = t\underline{i}+t^2\underline{j}$. Find the magnitude of the tangential and normal components of the acceleration.

****

$$\underline{r} = (t, t^2)$$

$$\underline{v} = \frac{d\underline{r}}{dt} = (1, 2t)$$

$$\underline{a} = \frac{d\underline{v}}{dt} = (0, 2)$$

$$\underline{V} = (1, 2t) = \frac{ds}{dt} \underline{t} \qquad \text{WHERE } \|\underline{t}\| = 1.$$

$$\text{So } \|V\| = (1+4t^2)^{1/2} = \frac{ds}{dt}.$$

THEREFORE

$$a_t = \frac{d^2s}{dt^2} = \frac{1}{2}(1+4t^2)^{-1/2}(8t) = 4t(1+4t^2)^{-1/2}$$

$$\|\underline{a}\|^2 = a_t^2 + a_n^2 \quad \text{so.}$$

$$4 = 8t^2(1+4t^2)^{-1} + a_n^2$$

$$\text{So } a_n^2 = 4 - \frac{8t^2}{(1+4t^2)} = \frac{4+8t^2}{1+4t^2}$$

$$a_n = 2\left(\frac{1+2t^2}{1+4t^2}\right)^{1/2}$$

━━━━━━━━━━━━━━━━━━━━━━━━━━━━━━━━━━ **1-37**

Calculate the tangential and normal components of acceleration for a particle moving in a helical path given by $\vec{r} = \hat{i}4\cos 3t + \hat{j}4\sin 3t + \hat{k}6t$.

****

$$\vec{v} = \frac{d\vec{r}}{dt} = -\hat{i}12\sin 3t + \hat{j}12\cos 3t + \hat{k}6$$

$$v = \sqrt{(-12\sin 3t)^2 + (12\cos 3t)^2 + (6)^2}$$

$$= \sqrt{144(\sin^2 3t + \cos^2 3t) + 36} = \sqrt{144(1) + 36}$$

$$= \sqrt{180} = 6\sqrt{5}$$

$$\hat{T} = \frac{\vec{v}}{v} = \frac{-\hat{i}12\sin 3t + \hat{j}12\cos 3t + \hat{k}6}{6\sqrt{5}}$$

$$= \frac{-\hat{i}2\sin 3t + \hat{j}2\cos 3t + \hat{k}}{\sqrt{5}}$$

$$\frac{d\hat{T}}{dt} = \frac{-\hat{i}6\cos 3t - \hat{j}6\sin 3t + 0}{\sqrt{5}}$$

$$\left|\frac{d\hat{T}}{dt}\right| = \sqrt{\left(-\frac{6\cos 3t}{\sqrt{5}}\right)^2 + \left(-\frac{6\sin 3t}{\sqrt{5}}\right)^2} = \sqrt{\frac{36}{5}(\cos^2 3t + \sin^2 3t)}$$

$$= \sqrt{\frac{36}{5}(1)} = \frac{6}{\sqrt{5}}$$

Curvature: $K = \frac{1}{v}\left|\frac{d\hat{T}}{dt}\right| = \frac{\frac{6}{\sqrt{5}}}{6\sqrt{5}} = \frac{1}{5}$

Tangential acceleration: $a_T = \frac{dv}{dt} = \frac{d}{dt}6\sqrt{5} = 0$

Normal acceleration: $a_N = Kv^2 = \frac{1}{5}(6\sqrt{5})^2 = 36$.

1-38 ■■

For $R(t) = t^2 \vec{i} + t \vec{j}$, find A_T and A_N, the tangential and normal components of acceleration.

****

$$\vec{R'(t)} = 2t\,\vec{i} + \vec{j}, \quad |\vec{R'(t)}| = \sqrt{4t^2+1}, \quad \text{hence} \quad A_T =$$

$$D_t\,|\vec{R'(t)}| = D_t(4t^2+1)^{\frac{1}{2}} = \tfrac{1}{2}(4t^2+1)^{-\frac{1}{2}}\cdot 8t = \frac{4t}{\sqrt{4t^2+1}}.$$

For the curvature,
$$\left.\begin{array}{ll} x = t^2 & y = t \\ x' = 2t & y' = 1 \\ x'' = 2 & y'' = 0 \end{array}\right\} \quad K = \frac{(2t)\cdot 0 - 2\cdot 1}{(4t^2+1)^{3/2}} = \frac{-2}{(4t^2+1)^{3/2}}.$$

Therefore, $A_N = K\cdot|\vec{R'(t)}|^2 = \frac{-2}{(4t^2+1)^{3/2}}\cdot(4t^2+1) = \frac{-2}{\sqrt{4t^2+1}}.$

PARAMETRIC EQUATIONS AND CURVATURE

━━━━━━━━━━━━━━━━━━━━━━━━━━━━━━━━━━━━━**1-39**

Sketch the given parametric equations where t is a real number.

a) $x = \dfrac{t}{2}$, $y = 1 - t$ b) $x = \dfrac{1}{2} - \dfrac{1}{2} t^2$, $y = t^2$

c) $x = \dfrac{1}{2} \cos^2 t$, $y = \sin^2 t$.

****

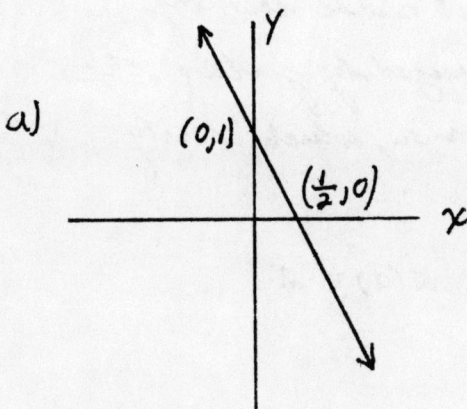

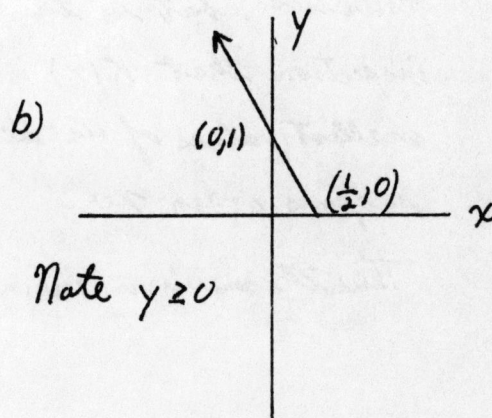

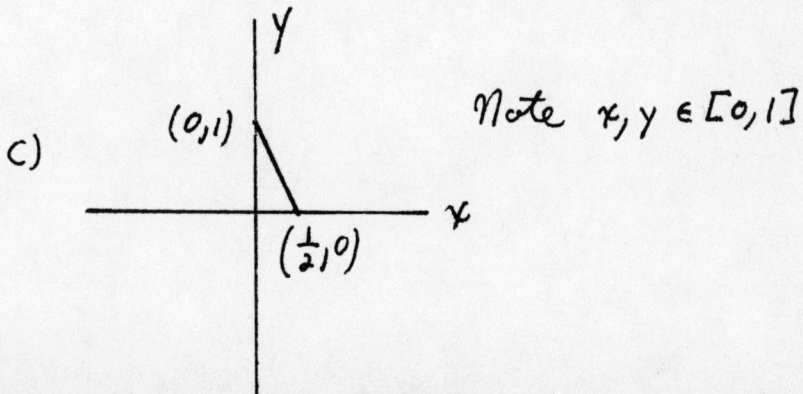

1-40 ■■■

What is the maximum curvature on the curve given by $y = x^2$?

****

Since $K(x) = \dfrac{\left|\dfrac{d^2y}{dx^2}\right|}{\left[1 + \left(\dfrac{dy}{dx}\right)^2\right]^{3/2}}$

we find $\dfrac{dy}{dx} = 2x$ and $\dfrac{d^2y}{dx^2} = 2$.

Thus $K(x) = \dfrac{2}{(1 + 4x^2)^{3/2}}$.

The maximum of this function can be found using derivatives, but in this case it can be seen by inspection that $K(x)$ is maximized by finding the smallest values of its denominator, which clearly happens when $x = 0$.

Thus the maximum curvature is $K(0) = 2$.

1-41

Consider the curve given by $x = t^2+3$ and $y = 2t^3-t$.

(a) Find $\dfrac{dy}{dx}$ at the point corresponding to t=2.

(b) Find $\dfrac{d^2y}{dx^2}$ at the point corresponding to t=2.

****

(a) $\left.\dfrac{dy}{dx}\right|_{t=2} = \left.\dfrac{dy/dt}{dx/dt}\right|_{t=2} = \left.\dfrac{6t^2-1}{2t}\right|_{t=2} = \dfrac{23}{4}$

(b) Let $y' = \dfrac{6t^2-1}{2t}$, then

$\dfrac{dy'}{dt} = \dfrac{(12t)(2t)-(6t^2-1)(2)}{4t^2} = \dfrac{6t^2+1}{2t^2}$

$\therefore \dfrac{d^2y}{dx^2} = \dfrac{dy'/dt}{dx/dt} = \dfrac{\left(\dfrac{6t^2+1}{2t^2}\right)}{2t}$

So, $\left.\dfrac{d^2y}{dx^2}\right|_{t=2} = \left.\dfrac{6t^2+1}{4t^3}\right|_{t=2} = \dfrac{25}{32}$

1-42 ■■■ ■■■ ■■■ ■■■ ■■■ ■■■ ■■■ ■■■ ■■■ ■■■ ■■■ ■■■ ■■■

Find the parametric equations of the line through the 2 points (1, 1)
and (2, -3).

****

We choose first to find the vector
equation of the line and then put
the line into parametric form.

Vector equation is $X = X_0 + tV$ where
X_0 is a point on the line, V is any
vector parallel to the line, X
is any variable point on the line
and t varies over the reals. Then

$$V = (2, -3) - (1, 1)$$
$$= (1, -4)$$

Therefore the vector equation is

$$X = (1, 1) + t(1, -4)$$

Hence $(x, y) = (1, 1) + t(1, -4)$

Using the elementary properties of
vectors we have

$$(x, y) = (1 + t, 1 - 4t); \text{ hence,}$$

the parametric equation of the
line is given by

$$x = 1 + t$$
$$y = 1 - 4t$$

1-43

If a circle rolls along a straight line without slipping, then a point on the circle traces a curve called a cycloid. If the circle has radius a, then the cycloid is given parametrically by:

$$x = at - a \sin t$$

$$y = a - a \cos t$$

a) Find $\dfrac{dy}{dx}$ for the cycloid.

b) Let M be the point on the circle which is furthest from the x-axis. Show that the line through M and the point P of the cycloid is tangent to the cycloid at P.

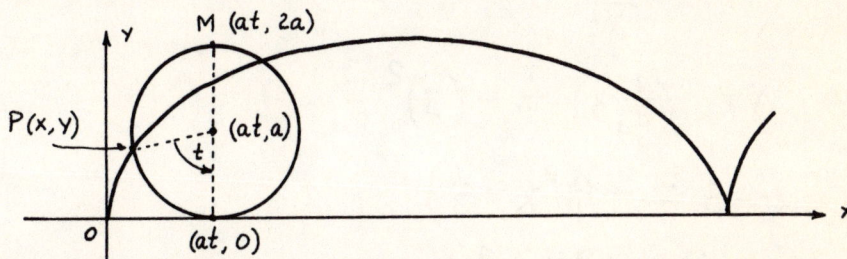

****

a) Now $\dfrac{dy}{dt} = a \sin t$ and $\dfrac{dx}{dt} = a - a \cos t$

So $\dfrac{dy}{dx} = \dfrac{dy}{dt} \cdot \dfrac{dt}{dx} = \dfrac{a \sin t}{a - a \cos t} = \underline{\underline{\dfrac{\sin t}{1 - \cos t}}}$

b) The coordinates of M are $(at, 2a)$ and since P is on the cycloid, its coordinates are $(at - a \sin t, a - a \cos t)$.

So, slope of $\overleftrightarrow{MP} = \dfrac{2a - (a - a \cos t)}{at - (at - a \sin t)} = \dfrac{1 + \cos t}{\sin t}$

Now if the slope of $\overleftrightarrow{MP}$ is equal to $\dfrac{dy}{dx}$ then $\overleftrightarrow{MP}$ is tangent to the cycloid.

$\dfrac{1 + \cos t}{\sin t} = \dfrac{1 + \cos t}{\sin t} \cdot \dfrac{1 - \cos t}{1 - \cos t} = \dfrac{\sin^2 t}{\sin t (1 - \cos t)} = \dfrac{\sin t}{1 - \cos t}$

As desired.

1-44 ■■■

Find an equation in x and y for the tangent line to the curve
$x = e^t$, $y = e^{-t}$ at the point ($\frac{1}{3}$, 3).

****

$$\frac{dx}{dt} = e^t \qquad\qquad \frac{dy}{dt} = -e^{-t}$$

$$SLOPE = \frac{dy}{dx} = \frac{\frac{dy}{dt}}{\frac{dx}{dt}} = \frac{-e^{-t}}{e^t} = -e^{-2t} = -\left(e^{-t}\right)^2$$

$$SLOPE \ AT \ \left(\tfrac{1}{3}, 3\right) = -(3)^2 = -9$$

$$y - 3 = -9\left(x - \tfrac{1}{3}\right)$$

$$y - 3 = -9x + 3$$

$$y = -9x + 6$$

2
VECTORS IN SPACE

VECTORS IN THREE - DIMENSIONAL SPACE

■■■ 2-1

Let A = (2,1,3),   B = (1,2,-2),   C = (-1,3,1).   Find an equation of the
plane through A, B, and C.

****

$\vec{AB} = \langle -1, 1, -5 \rangle$, $\vec{AC} = \langle -3, 2, -2 \rangle$. A normal vector

for the plane is $\vec{AB} \times \vec{AC} = \begin{vmatrix} \vec{i} & \vec{j} & \vec{k} \\ -1 & 1 & -5 \\ -3 & 2 & -2 \end{vmatrix} = \langle 8, 13, 1 \rangle$, so the

equation has the form $8x + 13y + z = d$. Putting

in A, $d = 16 + 13 + 3 = 32$, hence $8x + 13y + z = 32$.

2-2 ■■

Let $\vec{u} = (1,0,2)$ $\vec{v} = (1,-1,2)$ $\vec{w} = (1,2,k)$

For what value(s) of k (if any) will $\vec{w}$ lie in the plane of $\vec{u}$ and $\vec{v}$?

****

For $\vec{w}$ to lie in the plane of $\vec{u}$ and $\vec{v}$, $\vec{w}$ must be a linear combination of $\vec{u}$ and $\vec{v}$,

i.e. $\alpha \vec{u} + \beta \vec{v} = \vec{w}$

So: $\alpha(1,0,2) + \beta(1,-1,2) = (1,2,k)$

or: $\begin{cases} \alpha + \beta = 1 & (1) \\ -\beta = 2 & (2) \\ 2\alpha + 2\beta = k & (3) \end{cases}$

From equations (1) & (2) we get: $\beta = -2$, $\alpha = 3$

Hence, from (3) $k = 2 \cdot 3 + 2(-2) = 6 - 4 = 2$

So: $k = 2$

2-3 ■■

If α, β, γ are the direction angles of $\vec{A} = 2\vec{i} + 3\vec{j} + \sqrt{3}\vec{k}$, then $\cos\beta$ is (a) $3/(5 + \sqrt{3})$ (b) 3 (c) 3/5 (d) 3/4 (e) 3/7.

****

$|\vec{A}| = \sqrt{4 + 9 + 3} = \sqrt{16} = 4$, so $\frac{1}{|\vec{A}|}\vec{A} = \frac{1}{4}(2\vec{i} + 3\vec{j} + \sqrt{3}\vec{k})$

$= \frac{1}{2}\vec{i} + \frac{3}{4}\vec{j} + \frac{\sqrt{3}}{4}\vec{k}$. Therefore, $\cos\beta = \frac{3}{4}$ (the component of $\vec{j}$).

LINES IN SPACE

■■ **2-4**

Give parametric equations for the line through the points P(2,-1,-2) and Q(3,1,4).

****

A VECTOR PARALLEL TO THE LINE IS
$$\overrightarrow{PQ} = [3-2, \; 1-(-1), \; 4-(-2)] = [1, 2, 6].$$
TWO SETS OF PARAMETRIC EQUATIONS FOR THE LINE ARE:

$$\begin{array}{lll}
x = 2 + t & & x = 3 + t \\
y = -1 + 2t & \text{AND} & y = 1 + 2t \\
z = -2 + 6t & & z = 4 + 6t
\end{array}$$

NOTE: OTHER SETS OF PARAMETRIC EQUATIONS ALSO DESCRIBE THE LINE.   ANY POINT ON THE LINE AND ANY VECTOR PARALLEL TO THE LINE WILL DO.

■■ **2-5**

Let a = 2i + j - k and b = i + 2j + 3k. Find an equation of the line parallel to a + b and passing through the tip of b.

****

$$r(t) = i + 2j + 3k + t(a + b)$$

$$r(t) = i + 2j + 3k + t(3i + 3j + 2k)$$

$$r(t) = (1 + 3t)i + (2 + 3t)j + (3 + 2t)k$$

2-6 ▪▪▪▪▪▪▪▪▪▪▪▪▪▪▪▪▪▪▪▪▪▪▪▪▪▪▪▪▪▪▪▪▪▪▪▪▪▪

Find the cosine of the acute angle between the lines x = 4-4t, y = 3-t, z= 1+5t and x = 4-t, y = 3+2t, z = 1.

****

WRITE IN VECTOR NOTATION.

$$L_1: \quad \underline{x} = (4,3,1) + t(-4,-1,5) = \underline{x}_1 + t\,\underline{v}_1$$

$$L_2: \quad \underline{x} = (4,3,1) + t(-1,2,0) = \underline{x}_2 + t\,\underline{v}_2$$

COS of ANGLE BETWEEN L_1 and L_1 is COS of ANGLE BETWEEN $\underline{v}_1$ and $\underline{v}_2$ so

$$\cos\theta = \frac{\underline{v}_1 \cdot \underline{v}_2}{\|\underline{v}_1\| \, \|\underline{v}_2\|} = \frac{4-2+0}{\sqrt{16+1+25} \, \sqrt{1+4}} = \frac{2}{\sqrt{42}\sqrt{5}}$$

$$= \frac{2\sqrt{210}}{210} = \frac{\sqrt{210}}{105}$$

2-7 ▪▪▪▪▪▪▪▪▪▪▪▪▪▪▪▪▪▪▪▪▪▪▪▪▪▪▪▪▪▪▪▪▪▪▪▪▪▪

Find symmetric equations of the line passing through (2,-3,4) and parallel to the vector $\overrightarrow{AB}$ where A and B are the points (-2,1,1) and (0,2,3), respectively.

****

$$\overrightarrow{AB} = (0+2, 2-1, 3-1) = (2,1,2) = (a,b,c)$$

$$\therefore \quad \frac{x-x_0}{a} = \frac{y-y_0}{b} = \frac{z-z_0}{c} \quad \text{BECOMES}$$

$$\frac{x-2}{2} = \frac{y+3}{1} = \frac{z-4}{2}$$

■■ **2-8**

Show that the lines

$$x = 1 - t \qquad\qquad x = 2 + 5t$$
$$y = 4 + 4t \qquad \text{and} \qquad y = 1 + t$$
$$z = 3 + 2t \qquad\qquad z = -3t$$

are skew.

****

The first line is parallel to the vector $-\vec{i} + 4\vec{j} + 2\vec{k}$ while the second line is parallel to $5\vec{i} + \vec{j} - 3\vec{k}$. These vectors are not parallel since neither one is a scalar multiple of the other. So the lines are not parallel.

If the lines intersected at some point (x_0, y_0, z_0) then there would exist parameters t_1 and t_2 such that

$$x_0 = 1 - t_1 \qquad\qquad x_0 = 2 + 5t_2$$
$$y_0 = 4 + 4t_1 \qquad \text{and} \qquad y_0 = 1 + t_2$$
$$z_0 = 3 + 2t_1 \qquad\qquad z_0 = -3t_2$$

So $x_0 = 1 - t_1 = 2 + 5t_2 \implies t_1 + 5t_2 = -1$
and $y_0 = 4 + 4t_1 = 1 + t_2 \implies 4t_1 - t_2 = -3$

Solving these simultaneously gives $t_1 = {}^{16}/{-21}, \; t_2 = {}^{1}/{-21}$

But $z_0 = 3 + 2t_1 = -3t_2 \implies 2t_1 + 3t_2 = 3$

The values found above for t_1 and t_2 are not a solution for this equation, meaning that the lines do not intersect.

The lines are not parallel and they do not intersect, therefore they are skew.

2-9

Show that the lines

$$x = 2 - t \qquad\qquad x = 3 + t$$
$$y = 2 + t \qquad \text{and} \qquad y = 3t$$
$$z = 2t \qquad\qquad z = 5 - 4t$$

are skew.

****

Changing the parameter in the second line to s, if the lines intersected, we would have

$$2 - t = 3 + s$$
$$2 + t = 3s$$
$$2t = 5 - 4s.$$

Solving the first two equations simultaneously we obtain $s = \frac{1}{4}$ and $t = -\frac{5}{4}$, which violates the third equation. Thus the lines do not intersect. The first line is parallel to the vector $\langle -1, 1, 2 \rangle$, the second to $\langle 1, 3, -4 \rangle$, so the lines are not parallel. Thus they are skew.

■■■ **2-10**

Refer to Problem 2-9. Find the distance between the lines.

****

To find the distance between them we need a vector perpendicular to both of them, like

$$\langle 1, 3, -4 \rangle \times \langle -1, 1, 2 \rangle = \langle 10, 2, 4 \rangle, \text{ or}$$

$\langle 5, 1, 2 \rangle$. Normalizing, $\frac{1}{\sqrt{30}} \langle 5, 1, 2 \rangle$ is a unit vector perpendicular to both lines. The point $(3,0,5)$ is on the second line; $(2,2,0)$ on the first. Subtracting, $\langle 1, -2, 5 \rangle$ is a vector joining the lines. Projecting this onto the unit normal obtained above, $\langle 1, -2, 5 \rangle \cdot \frac{1}{\sqrt{30}} \langle 5, 1, 2 \rangle$, the distance between the lines is $\frac{13}{\sqrt{30}}$.

2-11 ▬▬▬▬▬▬▬▬▬▬▬▬▬▬▬▬▬▬▬▬▬▬▬▬▬▬▬▬▬▬▬▬▬▬▬

Compare the sets $\{(x,y,z) \mid z = 1 + 2x, \ (x,y) \ \varepsilon \ \{(x,y) \mid x + y = 3\}\}$ and $\{(x,y,z) \mid z = 7 - 2y, \ (x,y) \ \varepsilon \ \{(x,y) \mid x + y = 3\}\}$.

****

From the 1ST set we have $x + y = 3$ and $z = 1 + 2x$. Now let $x = t \Rightarrow \begin{cases} x = t \\ y = 3 - t \\ z = 1 + 2t \end{cases}$ where t is a real number. This is a line in parametric form. On the 2nd set we have that $x + y = 3$ and $z = 7 - 2y \Rightarrow z = 7 - 2(3 - x) \Rightarrow z = 7 - 6 + 2x = 1 + 2x$. therefore $x + y = 3$ and $z = 1 + 2x$ is also true in the 2nd set. These set are the same, each a different representation of a straight line.

━ **2-12**

Let 1 and 1' be two lines in space given by the equations:

$$1: \begin{cases} x = 3+t \\ y = 1-t \\ z = 2t \end{cases} \qquad 1': \begin{cases} x = -1+t \\ y = 2t \\ Z = 1+kt \end{cases}$$

Find all values of k (if any) for which:
(a) 1 and 1' are parallel
(b) 1 and 1' are perpendiculer

****

Let $\vec{u} = (1,-1,2)$ then $l \parallel \vec{u}$

Let $\vec{v} = (1, 2, k)$ then $l' \parallel \vec{v}$

a) For $l \parallel l'$, we need $\vec{u} \parallel \vec{v}$, i.e. $\vec{u} = \lambda \vec{v}$

$$\text{or } (1,-1,2) = \lambda(1,2,k)$$

hence: $\lambda = 1$ ←   Since these 2 equations are

$2\lambda = -1$ ←   inconsistent, there is <u>no</u>

$k\lambda = 2$ such λ, and hence no such k

Answer: <u>no</u> value of k for $l \parallel l'$

b) For $l \perp l'$, we need $\vec{u} \perp \vec{v}$, i.e. $\vec{u} \cdot \vec{v} = 0$

$\vec{u} \cdot \vec{v} = 1\cdot1 + (-1)\cdot2 + 2k = -1 + 2k = 0$

$\therefore l \perp l'$ when $k = \frac{1}{2}$

2-13 ■■

Find the equation of the line through the points (1, 1, −1) and (2, −1, 1).

****

The vector equation is given by
$$\vec{X} = \vec{X}_0 + t\vec{v}.$$

Choose $\vec{v}$ to be $(2, -1, 1) - (1, 1, -1) = (1, -2, 2)$
and $\vec{X}_0 = (1, 1, -1)$ so that
$$\vec{X} = (1, 1, -1) + t(1, -2, 2)$$
$$= (1+t, \; 1-2t, \; -1+2t)$$

Therefore the equation of the line
in 3 dimensional space is given
by
$$x = 1 + t$$
$$y = 1 - 2t$$
$$z = -1 + 2t$$

■■ **2-14**

Find symmetric and parametric equations of the line that goes through the
points P(1,2,4) and Q(3.-1.6) .

****

To write either equation we need any point the line
contains, and a vector parallel to the line. We will
use the point P for the point contained. Since P and Q
are both on the line, the vector $\overrightarrow{PQ}$ is parallel to
the line. $\overrightarrow{PQ} = (3-1)\vec{i} + (-1-2)\vec{j} + (6-4)\vec{k}$
$$= 2\vec{i} - 3\vec{j} + 2\vec{k}$$

Symmetric form: $\dfrac{x-1}{2} = \dfrac{y-2}{-3} = \dfrac{z-4}{2}$

Parametric form:
$$x = 1 + 2t$$
$$y = 2 - 3t$$
$$z = 4 + 2t$$

■■ **2-15**

Find an equation of the plane through the point P = (2,1,-4)   and perpendicular to the line x = 2 + 3t, y = 1 - 4t, z = 3 + 3t.

****

The vector $\langle 3, -4, 3 \rangle$ (coefficients of t) is parallel to

the line, hence normal to the plane. So the equation

has the form $3x - 4y + 3z = d$. To determine d,

put in P, $d = 6 - 4 - 12 = -10$. $3x - 4y + 3z = -10.$

2-16 ▬▬▬▬▬▬▬▬▬▬▬▬▬▬▬▬▬▬▬▬▬▬▬▬▬▬▬▬

Find the intersection point of the line $\vec{P} = (1,0,2) + (2,-2,1)t$ and the plane $3x + 4y + 6z = 7$.

****

$\vec{P}$: $x = 1 + 2t$, $y = -2t$, $z = 2 + t$

$3(1+2t) + 4(-2t) + 6(2+t) = 7$

$3 + 6t - 8t + 12 + 6t = 7 \rightarrow 4t = -8 \rightarrow t = -2$

$x = 1 + 2(-2) = -3,\ y = -2(-2) = 4,\ z = 2 + (-2) = 0$

ANS: $(-3, 4, 0)$

2-17 ▬▬▬▬▬▬▬▬▬▬▬▬▬▬▬▬▬▬▬▬▬▬▬▬▬▬▬▬

Find the equation of the line containing the points $(2,-2,3)$ and $(-5,-2,-3)$, in symmetric form.

****

Direction of line $= [2-(-5)]\,\vec{i} + [-2-(-2)]\,\vec{j} + [3-(-3)]\,\vec{k}$

$= 7\vec{i} + 6\vec{k}$

$\therefore$ Symmetric equations of the line passing through $(2,-2,3)$ with direction $7\vec{i} + 6\vec{k}$

are : $\dfrac{x-2}{7} = \dfrac{z-3}{6}$, $y = -2$

RECTANGULAR COORDINATES IN SPACE

■■■ **2-18**

Find the coordinates of the point halfway between the midpoints of the
vectors $a = 3i - 5j + k$ and $b = 5i + 3j + k$.

****

$$\frac{1}{2}\left(\frac{1}{2}(3,-5,1) + \frac{1}{2}(5,3,1)\right)$$

$$= \frac{1}{2}\left(\left(\frac{3}{2}, -\frac{5}{2}, \frac{1}{2}\right) + \left(\frac{5}{2}, \frac{3}{2}, \frac{1}{2}\right)\right)$$

$$= \frac{1}{2}\left(4, -1, 1\right) = \left(2, -\frac{1}{2}, \frac{1}{2}\right)$$

■■■ **2-19**

The center of the sphere $x^2 + y^2 + z^2 + 4x - 2y - 9z = 0$ is (a) (1,1,1)
(b) (0,0,0) (c) (4,-2,9) (d) (2,-1,-3) (e) (-2,1,3).

****

To see the center, put the equation in the form
$(x-a)^2 + (y-b)^2 + (z-c)^2 = r^2$, then (a,b,c) is the
center. Completing the squares,

$$x^2 + 4x + 4 + y^2 - 2y + 1 + z^2 - 6z + 9 = 14,$$

$$(x+2)^2 + (y-1)^2 + (z-3)^2 = 14,$$

so the center is $(-2, 1, 3)$.

2-20 ▬▬▬▬▬▬▬▬▬▬▬▬▬▬▬▬▬▬▬▬▬▬▬▬▬▬▬▬▬▬▬

Find the equation of the sphere, in standard form, one of whose diameters has (−5,2,9) and (3,−6,−1) as endpoints.

****

The center is the midpoint of the diameter.

$$\left(\frac{-5+3}{2},\ \frac{2+(-6)}{2},\ \frac{9+(-1)}{2}\right) = (-1, -2, 4)$$

The radius is the distance from the center to <u>any</u> point on the sphere. (We'll use (−5, 2, 9).)

$$r = \sqrt{(-5-(-1))^2 + (2-(-2))^2 + (9-4)^2} = \sqrt{16+16+25} = \sqrt{57}$$

so $r^2 = 57$.

Thus the equation is $(x+1)^2 + (y+2)^2 + (z-4)^2 = 57$

Simplifying: $x^2 + 2x + 1 + y^2 + 4y + 4 + z^2 - 8z + 16 = 57$

$$x^2 + 2x + y^2 + 4y + z^2 - 8z + 21 = 57$$

Answer: $x^2 + y^2 + z^2 + 2x + 4y - 8z - 36 = 0$

CYLINDRICAL AND SPHERICAL COORDINATES

■■■ **2-21**

(a) If $P = (1, 1, \sqrt{2})$ in rectangular coordinates, find spherical coordinates of P.

(b) If $Q = (1, \frac{\pi}{2}, 3)$ in cylindrical coordinates, find rectangular coordinates of Q.

****

(a) $\rho = \sqrt{x^2 + y^2 + z^2} = \sqrt{1 + 1 + 2} = \sqrt{4} = 2$;

$\tan \theta = y/x = 1$, $\quad \theta = \frac{\pi}{4}$;

$\cos \varphi = \frac{z}{\rho} = \frac{\sqrt{2}}{2}$, $\quad \varphi = \frac{\pi}{4}$;

$(\rho, \theta, \varphi) = \left(2, \frac{\pi}{4}, \frac{\pi}{4}\right)$

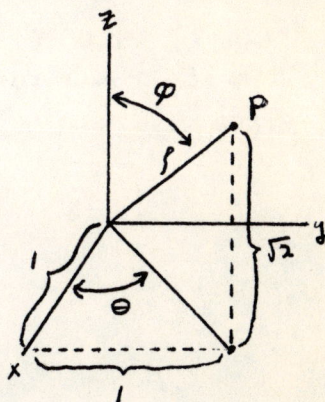

(b) $r = 1$, $\theta = \frac{\pi}{2} \implies (x, y) = (0, 1)$, so $(x, y, z) = (0, 1, 3)$

2-22 ■■

Sketch the graph of the spherical coordinate equation

$$\rho = 1 + \cos \varphi$$

****

Here is a table of values by intervals for ρ.

$$\varphi = 0: \quad \rho = 1 + \cos 0 = 1 + 1 = 2$$

$$0 \leq \varphi \leq \frac{\pi}{2}: \quad 2 \geq \rho \geq 1$$

$$\frac{\pi}{2} \leq \varphi \leq \pi: \quad 1 \geq \rho \geq 0$$

That is, as φ changes from 0 to $\frac{\pi}{2}$, ρ decreases from 2 to 1, and so on. This gives the trace in the x≥0 plane.

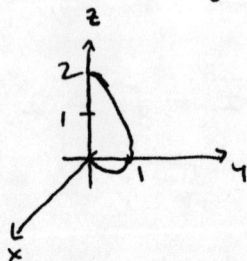

To get the complete graph, revolve this curve about the z axis, since θ is not restricted. The result is a three dimensional cardioid.

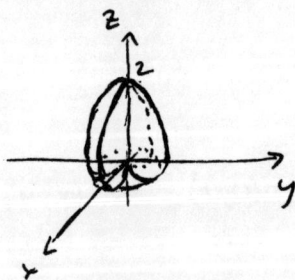

━━━━━━━━━━━━━━━━━━━━━━━━━━━━━━━━━━━━━━━ **2-23**

The following points are given in rectangular coordinates. Convert each one to cylindrical <u>and</u> spherical coordinates.

(a) $(1, -\sqrt{3}, -2)$

(b) $(0, -5, 0)$

****

(a) $\tan \theta = \dfrac{y}{x} = -\sqrt{3}$, and the pt. is below

the 4th quadrant of the xy-plane $\therefore \theta = \dfrac{-\pi}{3}$

$r = \sqrt{x^2 + y^2} = \sqrt{1+3} = 2$

$\therefore$ cyl. coord. are $\left(2, \dfrac{-\pi}{3}, -2\right)$

$\rho = \sqrt{r^2 + z^2} = \sqrt{4+4} = 2\sqrt{2}$

$\phi = \arccos\left(\dfrac{z}{\rho}\right) = \arccos\left(\dfrac{-1}{\sqrt{2}}\right) = \dfrac{3\pi}{4}$

$\therefore$ spherical coord. are $\left(2\sqrt{2}, \dfrac{-\pi}{3}, \dfrac{3\pi}{4}\right)$

(b) The pt. is in the xy-plane on the negative y-axis

$\therefore \theta = \dfrac{3\pi}{2}$, $\phi = \dfrac{\pi}{2}$, and $r = \rho = 5$

cyl. coord. are $\left(5, \dfrac{3\pi}{2}, 0\right)$

spherical coord. are $\left(5, \dfrac{3\pi}{2}, \dfrac{\pi}{2}\right)$

THE DOT PRODUCT

2-24 ■■■■■■■■■■■■■■■■■■■■■■■■■■■■■■■■■■■■■■■

Let $A = \langle 0,1,2 \rangle$, $B = \langle -1,-1,3 \rangle$, $C = \langle 2,4,-2 \rangle$ and $D = \langle 1,3,-3 \rangle$
Find: (a) $A \cdot D - B \cdot C$
          (b) $(A \cdot D)B - (B \cdot C)D$
          (c) $(2A + B) \cdot (2C - D)$

****

(a) $A \cdot D = (0)(1) + (1)(3) + (2)(-3) = 0 + 3 - 6 = -3$

$B \cdot C = (-1)(2) + (-1)(4) + (3)(-2) = -2 - 4 - 6 = -12$

$\therefore A \cdot D - B \cdot C = (-3) - (-12) = 9$

(b) From part (a), $A \cdot D = -3$ so $(A \cdot D)B = -3B$
$$= \langle 3, 3, -9 \rangle$$

also from part (a), $B \cdot C = -12$ so $(B \cdot C)D = -12D$
$$= \langle -12, -36, 36 \rangle$$

$\therefore (A \cdot D)B - (B \cdot C)D = \langle 3, 3, -9 \rangle - \langle -12, -36, 36 \rangle$
$$= \langle 15, 39, -45 \rangle$$

(c) $2A + B = 2\langle 0,1,2 \rangle + \langle -1,-1,3 \rangle = \langle -1, 1, 7 \rangle$
and
$2C - D = 2\langle 2,4,-2 \rangle - \langle 1,3,-3 \rangle = \langle 3, 5, -1 \rangle$

$\therefore (2A+B) \cdot (2C-D) = \langle -1, 1, 7 \rangle \cdot \langle 3, 5, -1 \rangle$
$$= (-1)(3) + (1)(5) + (7)(-1)$$
$$= -3 + 5 - 7 = -5$$

■■■**2-25**

Determine all values for a so that the vectors $\vec{x} = 2\vec{i} + \vec{j} + 2\vec{k}$ and $\vec{y} = \vec{i} + 2\vec{j} + a\vec{k}$ will form a $60°$ angle.

****

$$\cos 60° = \frac{1}{2} = \frac{\vec{x} \cdot \vec{y}}{|\vec{x}||\vec{y}|} = \frac{2+2+2a}{\sqrt{4+1+4}\sqrt{1+4+a^2}}$$

$$= \frac{4+2a}{3\sqrt{5+a^2}} = \frac{1}{2} \implies 8+4a = 3\sqrt{5+a^2}$$

$$\implies 64 + 64a + 16a^2 = 45 + 9a^2$$

$$7a^2 + 64a + 19 = 0$$

$$a = \frac{-64 \pm \sqrt{64^2 - 4(7)(19)}}{14}$$

$$= \frac{-64 \pm \sqrt{3564}}{14} \doteq \frac{-64 \pm 18\sqrt{11}}{14}$$

$$= \frac{-32 \pm 9\sqrt{11}}{7}$$

$$\therefore \quad a = \frac{-32 + 9\sqrt{11}}{7} \quad \text{and} \quad a = \frac{-32 - 9\sqrt{11}}{7}$$

are the two desired values.

2-26 ■■

Suppose $\vec{u}$ = <1,2,3> and $\vec{v}$ = <1,1,-2>. Find two vectors $\vec{a}$ and $\vec{b}$ such that $\vec{u} = \vec{a} + \vec{b}$, $\vec{a}$ is parallel to $\vec{v}$, and $\vec{b}$ is perpendicular to $\vec{v}$.

****

$$\vec{a} = \text{projection of } \vec{u} \text{ onto } \vec{v} = \left(\frac{\vec{u}\cdot\vec{v}}{\vec{v}\cdot\vec{v}}\right)\vec{v}$$

$$= \frac{-3}{6}\langle 1, 1, -2\rangle = \langle -\tfrac{1}{2}, -\tfrac{1}{2}, 1\rangle$$

$$\vec{b} = \vec{u} - \vec{a} = \langle 1, 2, 3\rangle - \langle -\tfrac{1}{2}, -\tfrac{1}{2}, 1\rangle$$

$$= \langle \tfrac{3}{2}, \tfrac{5}{2}, 2\rangle$$

(Check: $\vec{b}\cdot\vec{v} = 0$, so $\vec{b}\perp\vec{v}$)

2-27 ■■

Let a = i + 2j + 3k and b = 2i - j + k. Find $\|a\|$, a·b, and a x b.

****

$$\|a\| = \sqrt{1^2 + 2^2 + 3^2} = \sqrt{14}$$

$$a\cdot b = 1\cdot 2 + 2\cdot(-1) + 3\cdot 1 = 3$$

$$a\times b = (2\cdot 1 - 3(-1))\,i + (3\cdot 2 - 1\cdot 1)\,j + (1(-1) - 2\cdot 2)\,k$$

$$a\times b = 5i + 5j - 5k$$

-- **2-28**

Given the vectors: $\vec{A} = 3\vec{i} - 2\vec{j} + \vec{k}$
$\vec{B} = \vec{i} - 3\vec{j} + 5\vec{k}$
$\vec{C} = 2\vec{i} + \vec{j} - 4\vec{k}$

Do these vectors form a right triangle. Show why or why not.

****

First, we must check to see if the vectors do indeed form a triangle.

Since $\vec{B} + \vec{C} = 3\vec{i} - 2\vec{j} + \vec{k} = \vec{A}$, the vectors do form a triangle.

If they form a right triangle, the dot product of one pair of vectors must be 0.

Since $\vec{A} \cdot \vec{C} = (3)(2) + (-2)(1) + (1)(-4)$

$$= 0$$

$\boxed{\text{the vectors do form a right triangle.}}$

-- **2-29**

If $\vec{A} = 3\vec{i} - 2\vec{j} + \vec{k}$ and $\vec{B} = \vec{i} + 2\vec{j} - 3\vec{k}$, then $\vec{A} \cdot \vec{B}$ is (a) $2\sqrt{6}$
(b) $3\vec{i} - 4\vec{j} - 3\vec{k}$ (c) 10 (d) -4 (e) 4.

****

$$\vec{A} \cdot \vec{B} = a_1 b_1 + a_2 b_2 + a_3 b_3 = 3 - 4 - 3 = -4.$$

2-30 ■■

For which values of t are $\vec{a}(t)$ = <t+2, t, t> and $\vec{b}(t)$ = <t-2, t+1, t> orthogonal?

****

two vectors are orthogonal if their dot product is zero, thus we will solve $\vec{a}(t) \cdot \vec{b}(t) = 0$.

$$<t+2, t, t> \cdot <t-2, t+1, t> = (t+2)(t-2) + t(t+1) +$$

$$t^2 = t^2 - 4 + t^2 + t + t^2 = 3t^2 + t - 4 = 0 \Rightarrow$$

$$(3t+4)(t-1) = 0 \Rightarrow t = -\tfrac{4}{3}, 1.$$ Therefore the vectors $\vec{a}$ and $\vec{b}$ are orthogonal when t is either $-\tfrac{4}{3}$ or 1.

2-31 ■■

Let $\vec{A} = 3\vec{i} - \vec{j} + 2\vec{k}$ and $B = \vec{i} - 2\vec{j} + 3\vec{k}$. Find the scalar projection of $\vec{A}$ onto $\vec{B}$.

****

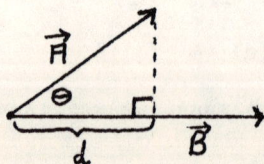

$$d = |\vec{A}| \cdot \cos\theta = \frac{|\vec{A}| \cdot |\vec{B}| \cdot \cos\theta}{|\vec{B}|} = \frac{\vec{A} \cdot \vec{B}}{|\vec{B}|} = \frac{3+2+6}{\sqrt{1+4+9}} = \frac{11}{\sqrt{14}}$$

■■ **2-32**

Find the projection of the vector [2,3,-5] along the vector [-1,1,2].

****

THE PROJECTION OF VECTOR $\vec{a}$ ALONG VECTOR $\vec{b}$ IS
A PRODUCT OF TWO FACTORS :

1) $\dfrac{\vec{a} \cdot \vec{b}}{|\vec{b}|}$, THE COMPONENT OF $\vec{a}$ IN THE
DIRECTION OF $\vec{b}$

2) $\dfrac{\vec{b}}{|\vec{b}|}$, THE UNIT VECTOR IN THE DIRECTION
OF $\vec{b}$.

SO THE PROJECTION OF $\vec{a}$ ALONG $\vec{b}$ IS GIVEN
BY $\dfrac{\vec{a} \cdot \vec{b}}{|\vec{b}|^2} \vec{b}$.

$$\frac{[2, 3, -5] \cdot [-1, 1, 2]}{(-1)^2 + 1^2 + 2^2} [-1, 1, 2] = \frac{2(-1) + 3(1) + (-5)(2)}{6} [-1, 1, 2]$$

$$= \frac{-9}{6} [-1, 1, 2]$$

$$= \left[\frac{3}{2}, -\frac{3}{2}, -3\right]$$

■■ **2-33**

Are the planes $3x - y + 5z = 13$ and $x + 7y - 2z = 4$ perpendicular to
each other?

****

A normal to the 1° plane is $n_1 = (3, -1, 5)$ and a
normal to the 2° plane is $n_2 = (1, 7, -2)$. The
given planes are $\perp$ to each other if $n_1 \perp n_2$,
i.e. if the dot product $n_1 \cdot n_2 = 0$.

But $n_1 \cdot n_2 = 3 - 7 - 10 \neq 0$, so the answer is $\boxed{No}$

2-34 ∎∎∎∎∎∎∎∎∎∎∎∎∎∎∎∎∎∎∎∎∎∎∎∎∎∎∎∎∎∎∎∎∎∎∎∎∎∎

Determine the vector component of x = (3, -4) which is parallel to
a = (5, 2).

****

By definition, we can write:

$$c\,\vec{a} = c\,(5,2)$$

such that

$$c = \frac{\vec{x} \cdot \vec{a}}{\vec{a} \cdot \vec{a}}$$

$$= \frac{(3, -4) \cdot (5, 2)}{(5, 2) \cdot (5, 2)}$$

$$= \frac{(3 \times 5) + (-4 \times 2)}{(5 \times 5) + (2 \times 2)}$$

$$= \frac{15 - 8}{25 + 4} = \frac{7}{29}$$

We then determine the needed vector component by writing:

$$\frac{7}{29}(5,2) = \left(\frac{7 \times 5}{29}, \frac{7 \times 2}{29} \right)$$

$$= \left(\frac{35}{29}, \frac{14}{29} \right) \quad \underline{Ans.}$$

■■■ 2-35

Find the angle between the lines $\frac{x-2}{1} = \frac{1-y}{3} = \frac{z-3}{1}$ and $\frac{x}{2} = \frac{y+3}{-1} = \frac{z-1}{2}$.

****

L_1:

$$\frac{X-2}{1} = t \rightarrow X = t+2$$
$$\frac{Y-1}{-3} = t \rightarrow Y = -3t+1$$
$$\frac{Z-3}{1} = t \rightarrow Z = t+3$$

$$\overrightarrow{L_1} = (2,1,3) + (1,-3,1)\,t$$
$$= \overrightarrow{P_1} + \overrightarrow{V_1}\,t$$

L_2:

$$\frac{X}{2} = t \rightarrow X = 2t$$
$$\frac{Y+3}{-1} = t \rightarrow Y = -3-t$$
$$\frac{Z-1}{2} = t \rightarrow Z = 1+2t$$

$$\overrightarrow{L_2} = (0,-3,1) + (2,-1,2)\,t$$
$$= \overrightarrow{P_2} + \overrightarrow{V_2}\,t$$

$$\overrightarrow{V_1} \cdot \overrightarrow{V_2} = |\overrightarrow{V_1}||\overrightarrow{V_2}|\cos\theta \quad \therefore \quad \theta = \cos^{-1} \frac{\overrightarrow{V_1}\cdot\overrightarrow{V_2}}{|\overrightarrow{V_1}||\overrightarrow{V_2}|}$$

$$= \cos^{-1} \frac{(1,-3,1)\cdot(2,-1,2)}{\sqrt{11}\ \sqrt{9}} = \cos^{-1} \frac{2+3+2}{3\sqrt{11}}$$

$$= \cos^{-1} \frac{7}{3\sqrt{11}} \approx 45.3° \text{ or } .7904 \text{ radians}$$

2-36 ■■■

Given vectors in R^3, A = i + j - 2k, B = j+k, C=3i-2j+k.

Find: 3A-B; $\|C\|$; B•C ; cosine of angle between B and C.

****

$$3A - B = 3(i + j - 2k) - (j + k) = (3i + 3j - 6k) - (j + k)$$
$$= 3i + 2j - 7k$$

$$\|C\| = \sqrt{3^2 + (-2)^2 + 1^2} = \sqrt{14}$$

$$B•C = (0i + j + k) • (3i - 2j + k) = 0(3) + 1(-2) + 1(1) = -1$$

$$\cos\theta = \frac{B•C}{\|B\|\,\|C\|} = \frac{-1}{\sqrt{2}\,\sqrt{14}} = \frac{-1}{\sqrt{28}} = \frac{-1}{2\sqrt{7}}$$

$B•C = -1$ (see above)

$\|C\| = \sqrt{14}$ (see above)

$\|B\| = \sqrt{0^2 + 1^2 + 1^2} = \sqrt{2}$

■■■ **2-37**

Let $\vec{P} = \vec{i} - 2\vec{j} + \vec{k}$ and $\vec{Q} = 2\vec{i} + \vec{j} + 2\vec{k}$. Find $\vec{P} - 2\vec{Q}$, $\vec{P} \cdot \vec{Q}$, and $\vec{P} \times \vec{Q}$

****

$$\vec{P} - 2\vec{Q} = \vec{i} - 2\vec{j} + \vec{k} - 2(2\vec{i} + \vec{j} + 2\vec{k})$$

$$= \vec{i} - 2\vec{j} + \vec{k} - 4\vec{i} - 2\vec{j} - 4\vec{k} = -3\vec{i} + 4\vec{j} - 8\vec{k}$$

$$\vec{P} \cdot \vec{Q} = (\vec{i} - 2\vec{j} + \vec{k}) \cdot (2\vec{i} + \vec{j} + 2\vec{k})$$

$$= (1)(2) + (-2)(1) + (1)(2) = 2 - 2 + 2 = 2$$

$$\vec{P} \times \vec{Q} = \begin{vmatrix} \vec{i} & \vec{j} & \vec{k} \\ 1 & -2 & 1 \\ 2 & 1 & 2 \end{vmatrix} = \vec{i} \begin{vmatrix} -2 & 1 \\ 1 & 2 \end{vmatrix} - \vec{j} \begin{vmatrix} 1 & 1 \\ 2 & 2 \end{vmatrix} + \vec{k} \begin{vmatrix} 1 & -2 \\ 2 & 1 \end{vmatrix}$$

$$= \vec{i}(-4-1) - \vec{j}(2-2) + \vec{k}(1+4) = -5\vec{i} + 5\vec{k}$$

■■■ **2-38**

Let $\vec{A} = (2,3,-4)$ and $\vec{B} = (5,-2,1)$ be vectors. Compute (a) $2\vec{A} + 3\vec{B}$, (b) The dot product $\vec{A} \cdot \vec{B}$ and (c) the vector product $\vec{A} \times \vec{B}$.

****

(a) $2\vec{A} + 3\vec{B} = 2(2,3,-4) + 3(5,-2,1) = (4,6,-8) + (15,-6,3)$

$$= (19, 0, -5)$$

(b) $\vec{A} \cdot \vec{B} = (2,3,-4) \cdot (5,-2,1) = 2 \cdot 5 + 3 \cdot (-2) - 4 \cdot 1 = 0$

(c) $\vec{A} \times \vec{B} = \begin{vmatrix} \vec{i} & \vec{j} & \vec{k} \\ 2 & 3 & -4 \\ 5 & -2 & 1 \end{vmatrix} = (3 \cdot 1 - (-4) \cdot (-2))\vec{i} - (2+20)\vec{j} - 19\vec{k}$

$$= (-5, -22, -19)$$

THE CROSS PRODUCT

2-39 ■■■

Consider the points P = (1,2,3), Q = (2,-1,0), and
R = (-1,4,1).

(a) Find an equation for the plane determined by P, Q, and R.

(b) Find the area of the triangle PQR.

****

(a) $\vec{PQ} = \langle 1, -3, -3 \rangle$ $\qquad$ $\vec{PR} = \langle -2, 2, -2 \rangle$

$$\vec{PQ} \times \vec{PR} = \begin{vmatrix} \vec{i} & \vec{j} & \vec{k} \\ 1 & -3 & -3 \\ -2 & 2 & -2 \end{vmatrix} = \langle 12, 8, -4 \rangle$$

$\therefore$ a normal to the plane is $\langle 3, 2, -1 \rangle$ and the form of the equation is

$3x + 2y - z = D$. Substituting P gives

$D = 3(1) + 2(2) - (3) = 4$

Equation is: $3x + 2y - z = 4$

(Check: Q and R also satisfy equation)

(b) Area $= \frac{1}{2} \| \vec{PQ} \times \vec{PR} \| = \frac{1}{2} \sqrt{224} = 2\sqrt{14}$

■■■ **2-40**

Find the area of the parallelogram PQRS with P(0,1,2), Q(4,1,1) and R(1,-1,3).

****

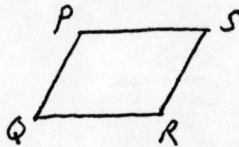

THE *AREA* OF THE PARALLELOGRAM IS THE LENGTH (MAGNITUDE) OF $\vec{QR} \times \vec{QP}$.

$$\vec{QR} = [1-4, -1-1, 3-1] = [-3, -2, 2]$$
$$\vec{QP} = [0-4, 1-1, 2-1] = [-4, 0, 1]$$

$$\vec{QR} \times \vec{QP} = \begin{vmatrix} \vec{i} & \vec{j} & \vec{k} \\ -3 & -2 & 2 \\ -4 & 0 & 1 \end{vmatrix} = \begin{vmatrix} -2 & 2 \\ 0 & 1 \end{vmatrix}\vec{i} - \begin{vmatrix} -3 & 2 \\ -4 & 1 \end{vmatrix}\vec{j} + \begin{vmatrix} -3 & -2 \\ -4 & 0 \end{vmatrix}\vec{k}$$

$$= (-2-0)\vec{i} - (-3+8)\vec{j} + (0-8)\vec{k}$$

$$= -2\vec{i} - 5\vec{j} - 8\vec{k} \quad OR \quad [-2, -5, -8]$$

$$|\vec{QR} \times \vec{QP}| = \sqrt{(-2)^2 + (-5)^2 + (-8)^2} = \sqrt{93}$$

2-41 ▬▬▬▬▬▬▬▬▬▬▬▬▬▬▬▬▬▬▬▬▬▬▬▬▬▬▬▬▬▬▬▬▬▬▬

If $\vec{a} = \vec{i} - 3\vec{j} + 2\vec{k}$ and $\vec{b} = 2\vec{i} - \vec{j} - 3\vec{k}$, find $\vec{a} \times \vec{b}$.

****

$$\vec{a} \times \vec{b} = \begin{vmatrix} \vec{i} & \vec{j} & \vec{k} \\ a_1 & a_2 & a_3 \\ b_1 & b_2 & b_3 \end{vmatrix}$$

$$\therefore \quad \vec{a} \times \vec{b} = \begin{vmatrix} \vec{i} & \vec{j} & \vec{k} \\ 1 & -3 & 2 \\ 2 & -1 & -3 \end{vmatrix}$$

EXPANDING BY THE TOP ROW GIVES

$$\vec{a} \times \vec{b} = \vec{i}\left[(-3)(-3) - 2(-1)\right] - \vec{j}\left[1(-3) - 2(2)\right] + \vec{k}\left[1(-1) - (-3)2\right]$$

$$= \vec{i}(9+2) - \vec{j}(-3-4) + \vec{k}(-1+6)$$

$$= 11\vec{i} + 7\vec{j} + 5\vec{k}$$

■■■ **2-42**

Find the equation of the plane containing the points (5,3,1), (1,8,4), and (-1,3,-2).

****

$$(5,3,1) - (1,8,4) = (4,-5,-3)$$

$$(5,3,1) - (-1,3,-2) = (6,0,3)$$

$$(4,-5,-3) \times (6,0,3) = (-15,-30,30)$$

$\therefore$ $(1, 2,-2)$ is a normal to the plane.

$$x + 2y - 2z = d$$

$$5 + 2(3) - 2(1) = d = 9$$

$\therefore$ $x + 2y - 2z = 9$ is the desired equation.

■■■ **2-43**

Let a and b be vectors. Under what conditions is a x b = b x a?
When is (2a) x (3b) = 6(a x b)?

****

$a \times b = b \times a$ WHEN (1) $a = b$
(2) $a = 0$ OR $b = 0$
(3) a AND b ARE PARALLEL
IN ALL CASES $a \times b = 0$

$(2a) \times (3b) = 6(a \times b)$ ALWAYS

2-44 ■■

Find the distance between the point (2,3,-1) and the line
$\frac{x+1}{3} = \frac{y-2}{5} = \frac{z+1}{-1}$.

****

First we write the line in parametric form

$$\begin{cases} x = -1 + 3t \\ y = 2 + 5t \\ z = -1 - t \end{cases}$$

$\overrightarrow{QP} = \langle 3,1,0 \rangle$ P(2,3,-1)

d

$\vec{\ell} = \langle 3,5,-1 \rangle$

θ

Q(-1,3,-1)

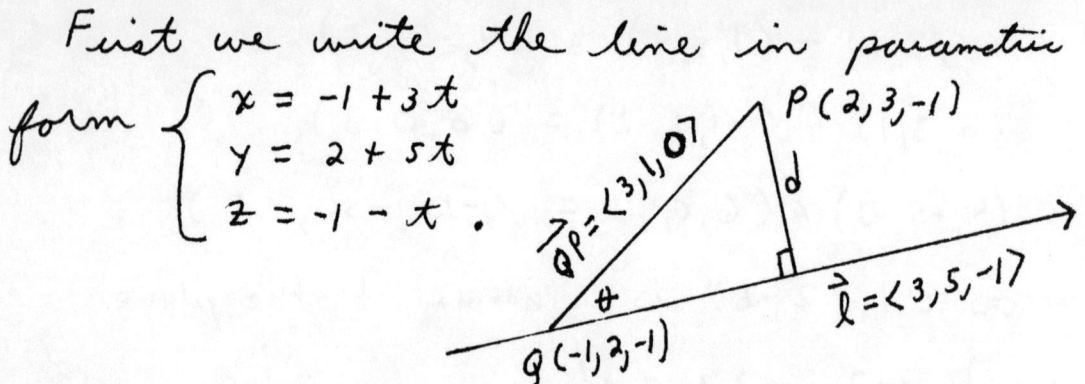

Now $\sin\theta = \dfrac{d}{|\overrightarrow{QP}|} \Rightarrow d = |\overrightarrow{QP}|\sin\theta \Rightarrow$

$d = |\overrightarrow{QP}|\dfrac{|\vec{\ell} \times \overrightarrow{QP}|}{|\vec{\ell}||\overrightarrow{QP}|} = \dfrac{|\vec{\ell} \times \overrightarrow{QP}|}{|\vec{\ell}|}$. Clearly, $|\vec{\ell}| =$

$\sqrt{(3)^2+(5)^2+(-1)^2} = \sqrt{9+25+1} = \sqrt{35}$ and $\vec{\ell} \times \overrightarrow{QP} =$

$\langle 1,-3,-12 \rangle$. Hence $|\vec{\ell} \times \overrightarrow{QP}| = \sqrt{1+9+144} = \sqrt{154}$.

Therefore, $d = \dfrac{\sqrt{154}}{\sqrt{35}} = \dfrac{\sqrt{7}\sqrt{22}}{\sqrt{7}\sqrt{5}} = \dfrac{\sqrt{22}\sqrt{5}}{5} = \dfrac{\sqrt{110}}{5}$.

2-45

a) Use the property of the cross product that

$$\| \vec{u} \times \vec{v} \| = \| \vec{u} \| \ \| \vec{v} \| \sin \theta$$

to derive a formula for the distance d from a point P to a line ℓ .

b) Use the formula obtained in part (a) to find the distance from the origin to the line through $(2,1,-4)$ and $(3,3,-2)$.

****

a) Let $\vec{u}$ be a vector from ℓ to P, $\vec{v}$ be a vector parallel to ℓ and θ be the angle between $\vec{v}$ and $\vec{u}$.

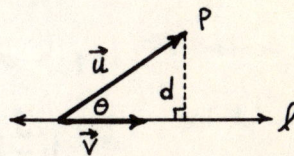

Note that $\quad d = \| \vec{u} \| \sin \theta$

But $\quad \| \vec{u} \times \vec{v} \| = \| \vec{u} \| \ \| \vec{v} \| \sin \theta$

So $\quad \| \vec{u} \times \vec{v} \| = \| \vec{v} \| d \implies d = \dfrac{\| \vec{u} \times \vec{v} \|}{\| \vec{v} \|}$

b) Since the origin has coordinates $(0,0,0)$ we

have $\quad \vec{u} = \langle 0-2, 0-1, 0-(-4) \rangle = \langle -2,-1,4 \rangle$

and $\quad \vec{v} = \langle 3-2, 3-1, -2-(-4) \rangle = \langle 1,2,2 \rangle$

Now $\quad \vec{u} \times \vec{v} = \begin{vmatrix} \vec{i} & \vec{j} & \vec{k} \\ -2 & -1 & 4 \\ 1 & 2 & 2 \end{vmatrix} = \langle -10, 8, -3 \rangle$

So $\quad \| \vec{u} \times \vec{v} \| = \sqrt{100 + 64 + 9} = \sqrt{173}$

And $\quad \| \vec{v} \| = \sqrt{1 + 4 + 4} = 3$

Thus $\quad d = \dfrac{\sqrt{173}}{3}$

2-46 ━━━

Given the vectors $\vec{u} = 2\vec{i} + \vec{j} - \vec{k}$
$\vec{w} = \vec{i} + \vec{j} + 4\vec{k}$

Find a vector of length 2 which is orthogonal to both $\vec{u}$ and $\vec{w}$.

****

Since $\vec{u} \times \vec{w}$ is orthogonal to both $\vec{u}, \vec{w}$

a vector of length 2 which is orthogonal to both $\vec{u}, \vec{w}$

would be $2 \left(\dfrac{\vec{u} \times \vec{w}}{|\vec{u} \times \vec{w}|} \right)$

$$\vec{u} \times \vec{w} = \begin{vmatrix} i & j & k \\ 2 & 1 & -1 \\ 1 & 1 & 4 \end{vmatrix} = (4 - -1)\vec{i} + (-1 - 8)\vec{j} + (2-1)\vec{k}$$

$$= 5\vec{i} + -9\vec{j} + \vec{k}$$

$$|\vec{u} \times \vec{w}| = \sqrt{25 + 81 + 1} = \sqrt{107}$$

Thus a vector of length 2 which is orthogonal to both $\vec{u}$ and $\vec{w}$ is

$$\boxed{\dfrac{10}{\sqrt{107}}\vec{i} - \dfrac{18}{\sqrt{107}}\vec{j} + \dfrac{2}{\sqrt{107}}\vec{k}}$$

■■ **2-47**

Let $\vec{u} \times \vec{v} = \vec{i} - \vec{j} + 2\vec{k}$. Find (if possible)

(a) $\vec{v} \times \vec{u}$ (b) $\vec{u} \cdot (\vec{u} \times \vec{v})$

(c) the area A of the parallelogram determined by $\vec{u}$ and $\vec{v}$

****

a) $\vec{v} \times \vec{u} = -(\vec{u} \times \vec{v}) = -\vec{i} + \vec{j} - 2\vec{k}$

b) Since $\vec{u} \perp (\vec{u} \times \vec{v})$, we have $\vec{u} \cdot (\vec{u} \times \vec{v}) = 0$

c) $A = \| \vec{u} \times \vec{v} \| = \sqrt{1^2 + 1^2 + 2^2} = \sqrt{6}$

■■ **2-48**

Let $A = \langle 3,0,-1 \rangle$, $B = \langle 4,1,-1 \rangle$, $C = \langle 2,1,0 \rangle$ and
   $D = \langle -1,-2,-2 \rangle$. Find: (a) $A \times B$ (b) $C \times D$

****

(a)
$$A \times B = \begin{vmatrix} i & j & k \\ 3 & 0 & -1 \\ 4 & 1 & -1 \end{vmatrix} = \begin{vmatrix} 0 & -1 \\ 1 & -1 \end{vmatrix} i - \begin{vmatrix} 3 & -1 \\ 4 & -1 \end{vmatrix} j + \begin{vmatrix} 3 & 0 \\ 4 & 1 \end{vmatrix} k$$

$$= \left[(0)(-1) - (1)(-1) \right] i - \left[(3)(-1) - (4)(-1) \right] j + \left[(3)(1) - (4)(0) \right] k$$

$$= i - j + 3k = \langle 1, -1, 3 \rangle.$$

(b)
$$C \times D = \begin{vmatrix} i & j & k \\ 2 & 1 & 0 \\ -1 & -2 & -2 \end{vmatrix} = \begin{vmatrix} 1 & 0 \\ -2 & -2 \end{vmatrix} i - \begin{vmatrix} 2 & 0 \\ -1 & -2 \end{vmatrix} j + \begin{vmatrix} 2 & 1 \\ -1 & -2 \end{vmatrix} k$$

$$= \left[(1)(-2) - (-2)(0) \right] i - \left[(2)(-2) - (-1)(0) \right] j + \left[(2)(-2) - (-1)(1) \right] k$$

$$= -2i + 4j - 3k = \langle -2, 4, -3 \rangle$$

2-49 ■■■

Determine an equation of the plane containing the following points :
(2, 0, 2), (4, 3, 1), and (0, 1, 3).

****

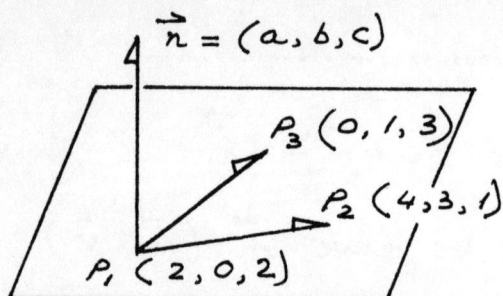

In the accompanying illustration, vectors $\vec{P_1P_2}$ and $\vec{P_1P_3}$ are shown lying in the plane and a unit vector $\vec{n}$ perpendicular to both $\vec{P_1P_2}$ & $\vec{P_1P_3}$.

As we need to find equations involving a, b, c to establish $\vec{n}$, we write:

$$\vec{P_1P_2} = (4-2, 3-0, 1-2) = (2, 3, -1)$$

and $$\vec{P_1P_3} = (0-2, 1-0, 3-2) = (-2, 1, -1)$$

As $\vec{n}$ is perpendicular to $\vec{P_1P_2}$ & $\vec{P_1P_3}$, we write:

$$(2, 3, -1) \cdot (a, b, c) = 0$$

and $$(-2, 1, -1) \cdot (a, b, c) = 0$$

Hence, a, b, and c has to satisfy the system of equations :

$$2a + 3b - c = 0$$
and $$-2a + b - c = 0$$

OR
$$2a + 3b = c \quad \ldots\ldots (1)$$
and $$-2a + b = c \quad \ldots\ldots (2)$$

Solving for a and b in simultaneous Eqs. (1) and (2) in terms of c we get:

$$a = -\frac{1}{4}c \quad \text{and} \quad b = \frac{1}{2}c$$

Now, to determine any one of the infinite number of vectors which are perpendicular to the plane, we choose a nonzero value for c and let c = 4.

Hence, we get :

$$a = -\frac{1}{4}(4) = -1 \quad \text{and} \quad b = \frac{1}{2}(4) = 2$$

Therefore, $n = (a, b, c) = (-1, 2, 4)$

By using P_1 (2, 0, 2) as the fixed point in the standard equation of the plane, we get:

$$-1(x-2) + 2(y-0) + 4(z-2) = 0$$

or $\quad -x + 2 + 2y - 0 + 4z - 8 \quad = 0$

$\quad\quad -x + 2y + 4z - 6 = 0 \quad$ <u>Ans.</u>

■■ **2-50**

Find the indicated area

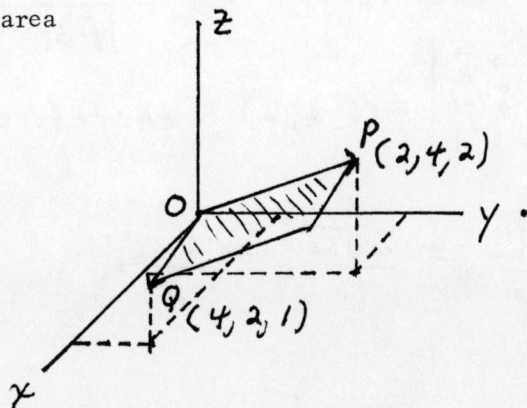

****

The area of the indicated region is

$$\left| \overrightarrow{OQ} \times \overrightarrow{OP} \right| = \left| \langle 2, 4, 2 \rangle \times \langle 4, 2, 1 \rangle \right| = \left| \langle 0, 6, -12 \rangle \right| =$$

$$\sqrt{(0)^2 + (6)^2 + 12^2} = \sqrt{36 + 144} = \sqrt{180} = \sqrt{9}\sqrt{4}\sqrt{5} =$$

$$3 \cdot 2\sqrt{5} = 6\sqrt{5} \text{ square units.}$$

2-51 ▰▰▰▰▰▰▰▰▰▰▰▰▰▰▰▰▰▰▰▰▰▰▰▰▰▰▰▰▰▰▰▰▰

Let P = (1,3,2) and let L be the line with parametric equations x = 2 - t, y = -1 + 2t, z = 3 + t. Use the vector cross-product to find the distance from P to L.

****

Letting $t=0$ and $t=1$, points $A = (2,-1,3)$ and $B=(1,1,4)$ of L are obtained. $\vec{AP} = \langle -1, 4, -1 \rangle$, $\vec{AB} = \langle -1, 2, 1 \rangle$

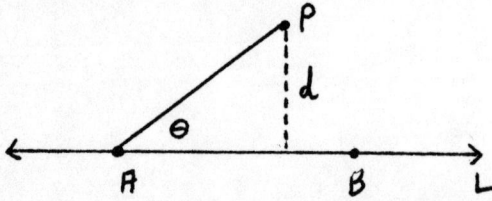

$$d = |\vec{AP}| \sin \theta$$

$$= \frac{|\vec{AP}| \cdot |\vec{AB}| \cdot \sin \theta}{|\vec{AB}|}$$

$$= \frac{|\vec{AP} \times \vec{AB}|}{|\vec{AB}|} .$$

$$\vec{AP} \times \vec{AB} = \begin{vmatrix} \vec{\imath} & \vec{\jmath} & \vec{k} \\ -1 & 4 & -1 \\ -1 & 2 & 1 \end{vmatrix} = \langle 6, 2, 2 \rangle, \text{ therefore } d = \frac{\sqrt{36+4+4}}{\sqrt{1+4+1}}$$

$$= \frac{\sqrt{44}}{\sqrt{6}} = \frac{2\sqrt{11}}{\sqrt{6}} = \frac{2\sqrt{66}}{6} = \frac{\sqrt{66}}{3} .$$

■■■ **2-52**

Find the area of the triangle defined by (1,0,2), (2,1,-3), and (3,-2,0).

****

$$A = (1,0,2) \qquad B = (2,1,-3) \qquad C = (3,-2,0)$$

$$\vec{AB} = (1,1,-5) \qquad \vec{AC} = (2,-2,-2)$$

$$\vec{AB} \times \vec{AC} = \begin{vmatrix} i & j & k \\ 1 & 1 & -5 \\ 2 & -2 & -2 \end{vmatrix} = (-2-10)\vec{i} - (-2+10)\vec{j} + (-2-2)\vec{k}$$

$$= (-12, -8, -4)$$

$$A_\Delta = \frac{1}{2} \left| \vec{AB} \times \vec{AC} \right| = \frac{1}{2}\sqrt{(-12)^2 + (-8)^2 + (-4)^2} = \frac{1}{2}\sqrt{224}$$

$$= 2\sqrt{14} \text{ units}^2$$

■■■ **2-53**

For vectors $\vec{A}$ and $\vec{B}$, $|\vec{A} \times \vec{B}| = |\vec{A}| \cdot |\vec{B}|$ provided (a) $\vec{A} = \vec{B}$ (b) $\vec{A}$ and $\vec{B}$ are parallel (c) $\vec{A}$ and $\vec{B}$ are not parallel (d) $\vec{A}$ and $\vec{B}$ are perpendicular (e) $\vec{A}$ and $\vec{B}$ are not perpendicular.

****

Since $|\vec{A} \times \vec{B}| = |\vec{A}| \cdot |\vec{B}| \sin\theta$, where θ is the angle between $\vec{A}$ and $\vec{B}$, we see that $|\vec{A} \times \vec{B}| = |\vec{A}| \cdot |\vec{B}|$ provided $\sin\theta = 1$, ie. $\theta = 90°$. In other words, when $\vec{A}$ and $\vec{B}$ are perpendicular.

2-54

Given U = 3i - 2j + k and V = i + j - 2k.

Find UxV and show that UxV is orthogonal to both U and V.

****

$$U \times V = \begin{vmatrix} i & j & k \\ 3 & -2 & 1 \\ 1 & 1 & -2 \end{vmatrix} = i \begin{vmatrix} -2 & 1 \\ 1 & -2 \end{vmatrix} - j \begin{vmatrix} 3 & 1 \\ 1 & -2 \end{vmatrix} + k \begin{vmatrix} 3 & -2 \\ 1 & 1 \end{vmatrix}$$

$$= 3i + 7j + 5k$$

Show $(U \times V) \cdot U = 0$

$(U \times V) \cdot U = (3i + 7j + 5k) \cdot (3i - 2j + k) = 9 - 14 + 5 = 0$

∴ U×V is orthogonal to U.

Show $(U \times V) \cdot V = 0$

$(U \times V) \cdot V = (3i + 7j + 5k) \cdot (i + j - 2k) = 3 + 7 - 10 = 0$

∴ U×V is orthogonal to V.

2-55

Find the normal vector to the plane which passes through the points (1,0,0), (0,0,1) and (4,3,-2).

****

A vector connecting any pair of these points will lie in the plane. The cross-product of 2 vectors in the plane will give a normal vector to the plane.

Choosing 2 pairs of points arbitrarily :

(1,0,0) and (0,0,1)   give the vector $(0,0,1) - (1,0,0) = (-1,0,1) = v_1$

(0,0,1) and (4,3,-2) give the vector $(4,3,-2) - (0,0,1) = (4,3,-3) = v_2$

$$n = v_1 \times v_2 = \begin{pmatrix} i & j & k \\ -1 & 0 & 1 \\ 4 & 3 & -3 \end{pmatrix} = i(0-3) - j(3-4) + k(-3-0)$$

$$\boxed{n = (-3, 1, -3)}$$

$\uparrow$ N.B. the sign here

2-56

If vector $\vec{A} = 3\vec{i} + 4\vec{j} - 2\vec{k}$, and vector $\vec{B} = 2\vec{i} - \vec{j} + 5\vec{k}$, find a vector perpendicular to both vectors.

****

A vector perpendicular to both $\vec{A}$ and $\vec{B}$ would be $\vec{A \times B}$.

$$\vec{A \times B} = \begin{vmatrix} \vec{i} & \vec{j} & \vec{k} \\ 3 & 4 & -2 \\ 2 & -1 & 5 \end{vmatrix} = \vec{i}\begin{vmatrix} 4 & -2 \\ -1 & 5 \end{vmatrix} - \vec{j}\begin{vmatrix} 3 & -2 \\ 2 & 5 \end{vmatrix} + \vec{k}\begin{vmatrix} 3 & 4 \\ 2 & -1 \end{vmatrix}$$

$$= \vec{i}(20 - 2) - \vec{j}(15 + 4) + \vec{k}(-3 - 8) = 18\vec{i} - 19\vec{j} - 11\vec{k}$$

Any scalar multiple of this vector would also be perpendicular to both $\vec{A}$ and $\vec{B}$.

QUADRIC SURFACES

2-57 ▬▬▬▬▬▬▬▬▬▬▬▬▬▬▬▬▬▬▬▬▬▬▬▬▬▬▬▬▬▬▬▬▬▬▬▬

Let S be the quadric surface given by:

$$x^2 + 2y^2 + 2x - z = 0$$

(a) What kind of surface is S?

(b) What are the traces of S in each of the three coordinate planes.

****

a) $x^2 + 2y^2 + 2x - z = 0$ Completing the square in x gives:

$$(x^2 + 2x + 4) + 2y^2 - z = 4$$

or: $z + 4 = (x+2)^2 + 2y^2$

So, S is an elliptic paraboloid.

b) xy-plane: $z = 0$ $(x+2)^2 + 2y^2 = 4$ ellipse.

yz-plane: $x = 0$ $z + 4 = 2y^2$: parabola

xz-plane: $y = 0$ $z + 4 = (x+2)^2$: parabola

2-58

Sketch the graphs of the following in 3-space. Name each surface.

(a) $y^2 + z = 4$ (b) $z = x^2 + y^2$ (c) $x^2/4 + y^2 - z^2 = 1$

****

(a) Since x is not involved, the graph is a cylinder whose intersection with the yz plane is the parabola $y^2 + z = 4$, hence a parabolic cylinder.

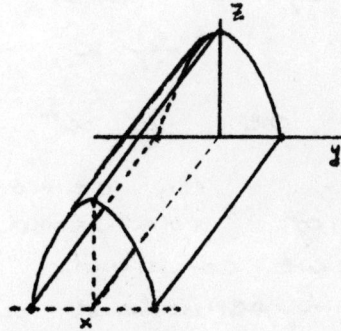

(b) Due to the $x^2 + y^2$ term, it's the paraboloid of revolution obtained by revolving $z = y^2$ about the z axis.

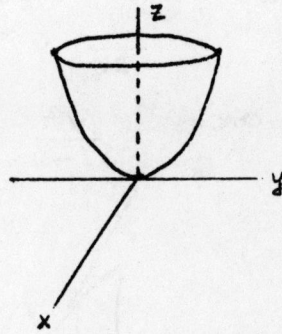

(c) By putting one variable equal to a constant, it is seen that cross-sections parallel to the xz and yz planes are hyperbolas, while cross-sections parallel to the xy plane are ellipses, hence it's an elliptic hyperboloid (of one sheet).

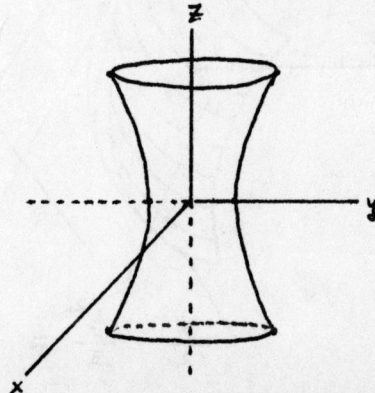

2-59 ■■

Identify the surface given by the following equations and sketch the graph given by any one of the equations :

(a) $6x^2 - 8y^2 + 3z^2 = 24$

(b) $3y^2 + 12z^2 - 36x = 0$

(c) $x^2 + z^2 - 8x = 0$

****

(a) We divide both sides of the given equation by 24 and write :

$$\frac{6}{24}x^2 - \frac{8}{24}y^2 + \frac{3}{24}z^2 = \frac{24}{24}$$

or $$\frac{1}{4}x^2 - \frac{1}{3}y^2 + \frac{1}{8}z^2 = 1$$

By comparing the above with the list of "standard equations" for quadric surfaces, we conclude that the surface generated is a hyperboloid of one sheet whose axis is the y-axis.

(b) We divide both sides of the given equation by 36 and write :

$$\frac{3}{36}y^2 + \frac{12}{36}z^2 - \frac{36}{36}x = 0$$

or $$\frac{y^2}{12} + \frac{z^2}{3} = x$$

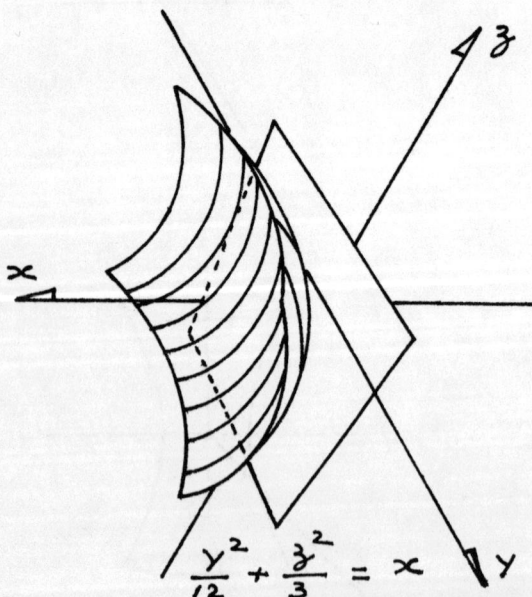

$$\frac{y^2}{12} + \frac{z^2}{3} = x$$

By comparison with the standard list for quadric surfaces, we conclude that the surface generated is an elliptic paraboloid with vertex at the origin whose axis is the x-axis.

The resulting graph is shown alongside.

(c) We note from the given equation that the y variable does not exist. Hence, we conclude that the surface is a cylinder. For completing the square in x, we rearrange the given equation and write:

$$x^2 - 8x + z^2 = 0$$
or $$(x - 4)^2 + z^2 = 16$$

By comparison, we conclude that it is the equation of a circular cylinder with the center axis parallel to the y-axis passing through (4,0,0) whose radius = 4. We also note that its graph is perpendicular to the xz - plane.

━━━━━━━━━━━━━━━━━━━━━━━━━━━━━━━━━━━━━ **2-60**

Which of the following is not a quadric surface? (a) $x^2 + z = 1$
(b) $z = x^2 + y^2$ (c) $y = x^3 + z$ (d) $z = x^2 - y^2$ (e) $x^2 + y^2 + z^2 = 1$.

****

The term "quadric surface" refers to the general quadratic equation in x, y, z:

$$Ax^2 + By^2 + Cz^2 + Dxy + Exz + Fyz + Gx + Hy + Iz + J = 0,$$

where A — J are constants. Hence $y = x^3 + z$ is not a quadric surface because of the x^3 term.

CURVES IN SPACE

2-61

Find the curvature of $\vec{r}(t) = \langle t, t^2, t^3 \rangle$ at $t = 2$.

****

The formula for curvature is

$$k = \frac{\|\vec{v} \times \vec{a}\|}{\|\vec{v}\|^3}.$$

Now $\vec{v}(t) = \dfrac{d\vec{r}}{dt} = \langle 1, 2t, 3t^2 \rangle$,

$\vec{v}(2) = \langle 1, 4, 12 \rangle$

$\vec{a}(t) = \dfrac{d\vec{v}}{dt} = \langle 0, 2, 6t \rangle$,

$\vec{a}(2) = \langle 0, 2, 12 \rangle$.

Since $\|\vec{v}(2)\| = \sqrt{1 + 16 + 144} = \sqrt{161}$ and

$\vec{v}(2) \times \vec{a}(2) = \langle 24, -12, 2 \rangle$,

$$k = \frac{\sqrt{724}}{(161)^{3/2}} = \frac{2\sqrt{181}}{161\sqrt{161}}.$$

■■■**2-62**

Consider the curve in the xy-plane defined parametrically by $x = t^3 - 3t$, $y = t^2$, $z = 0$.

(a) Find the slopes of the tangent lines to the curve at $t = \pm\ 3$.

(b) Find the unit tangent vector at $t = -1$.

(c) Find all points where the curve has vertical or horizontal tangents.

(d) Sketch a rough graph of the curve.

****

(a) $dx/dt = 3t^2 - 3$, $dy/dt = 2t$

$dy/dx = \dfrac{2t}{3(t^2-1)}$

∴ slope of tangent line at $t = 3$ is $1/4$
slope of tangent line at $t = -3$ is $-1/4$

(b) tangent vector at $t = -1$ is $0\vec{i} - 2\vec{j}$, or $-2\vec{j}$
unit tangent vector at $t = -1$ is $-\vec{j}$

(c) $dy/dt = 0$ at $t = 0 \Rightarrow$ horizontal tangent
at $(0,0)$

$dx/dt = 0$ at $t = \pm 1 \Rightarrow$ vertical tangents
at $(2,1)$ and $(-2,1)$

(d)

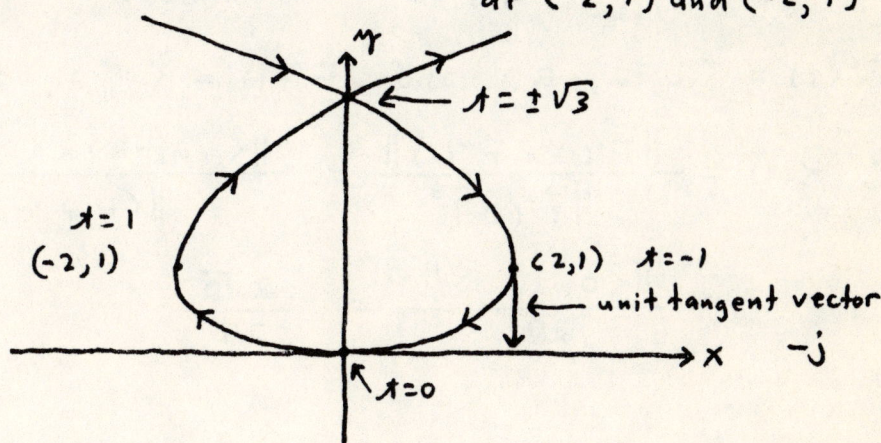

2-63 ■■

Suppose C is the curve given by the vector valued function
$\vec{r}(t) = < t, t^2, 1-t^2 >$. Find the unit tangent vector, the
unit normal vector, and the curvature of C at the point t=1.

****

$$\vec{r}'(t) = \langle 1, 2t, -2t \rangle \qquad \vec{T}(t) = \vec{r}'(t) / \| \vec{r}'(t) \|$$

$$\vec{T}(t) = \frac{1}{\sqrt{1+8t^2}} \langle 1, 2t, -2t \rangle$$

$$\therefore \ \vec{T}(1) = \frac{1}{3} \langle 1, 2, -2 \rangle$$

$$\vec{T}'(t) = \frac{-8t}{(1+8t^2)^{3/2}} \langle 1, 2t, -2t \rangle + \frac{1}{\sqrt{1+8t^2}} \langle 0, 2, -2 \rangle$$

$$\vec{T}'(1) = \frac{-8}{27} \langle 1, 2, -2 \rangle + \frac{1}{3} \langle 0, 2, -2 \rangle = \frac{1}{27} \langle -8, 2, -2 \rangle$$

So, $\langle -4, 1, -1 \rangle$ is in the direction of $\vec{N}(1)$.

$$\therefore \ \vec{N}(1) = \frac{1}{\sqrt{18}} \langle -4, 1, -1 \rangle \qquad (\text{Check: } \vec{T} \cdot \vec{N} = 0)$$

$$\vec{r}''(t) = \langle 0, 2, -2 \rangle \quad \text{and} \quad \vec{r}'(1) = \langle 1, 2, -2 \rangle$$

$$\therefore \ K(1) = \frac{\| \vec{r}'(1) \times \vec{r}''(1) \|}{\| \vec{r}'(1) \|^3} = \frac{\| \langle 1, 2, -2 \rangle \times \langle 0, 2, -2 \rangle \|}{\| \langle 1, 2, -2 \rangle \|^3}$$

$$= \frac{\| \langle 0, 2, 2 \rangle \|}{27} = \frac{2\sqrt{2}}{27}$$

■■ **2-64**

Suppose the curve C is given by the following vector function:
f(t)= (sin t, cos t, t^2), where t is between 0 and 1.
(a) Find the velocity vector at t= $\frac{\pi}{4}$.   (b) Find the acceleration
vector at t= $\frac{\pi}{4}$.

****

(a) The velocity vector is given by
$$v(t) = f'(t) = (\cos t, -\sin t, 2t); \text{ whence,}$$
$$V\left(\tfrac{\pi}{4}\right) = f'\left(\tfrac{\pi}{4}\right) = \left(\cos\tfrac{\pi}{4}, -\sin\tfrac{\pi}{4}, \tfrac{\pi}{2}\right) -$$
$$= \left(\tfrac{\sqrt{2}}{2}, -\tfrac{\sqrt{2}}{2}, \tfrac{\pi}{2}\right)$$

(b) The acceleration vector is given
by $a(t) = V'(t) = f''(t) = (-\sin t, -\cos t, 2);$
$$\text{whence, } a\left(\tfrac{\pi}{4}\right) = \left(-\sin\tfrac{\pi}{4}, -\cos\tfrac{\pi}{4}, 2\right)$$
$$= \left(-\tfrac{\sqrt{2}}{2}, -\tfrac{\sqrt{2}}{2}, 2\right)$$

■■■ **2-65**

Find the length of arc along the curve $\vec{x}(t)$ = (5t, sin t, cos t) from
t = 0 to t = 2.

****

$$\frac{d\vec{x}(t)}{dt} = (5, \cos t, -\sin t)$$

$$s = \int_0^2 \left| \frac{d\vec{x}(t)}{dt} \right| dt = \int_0^2 \sqrt{5^2 + (\cos t)^2 + (-\sin t)^2}\; dt$$

$$= \int_0^2 \sqrt{25+1}\; dt = \int_0^2 \sqrt{26}\; dt = 2\sqrt{26}$$

2-66 ■■■

Let C be the curve given by:

$$x(t) = e^t \qquad y(t) = t^2 \qquad z(t) = te^t$$

Find the point(s) P on C (if any) at which the tangent line to C is parallel to the xy-plane.

****

the tangent vector to C is given by:

$$\vec{T} = (f_x, f_y, f_z) = (e^t, 2t, te^t + e^t)$$

For $\vec{T} \parallel$ xy-plane, we need $f_z = 0$

$$te^t + e^t = (t+1) e^t = 0$$

Since $e^t > 0$ for all t, $f_z = 0$ only when $t = -1$
Hence, the tangent to C is parallel to the xy-plane at:

$$(x(-1), y(-1), z(-1)) = (e^{-1}, 1, -e^{-1})$$

2-67 ■■■

Find the length of the circular helix described by x = 2cos t, y = 2sin t, z = $\sqrt{5}$ t, $0 \leq t \leq 2\pi$.

****

$$x'(t) = -2\sin t, \quad y'(t) = 2\cos t, \quad z'(t) = \sqrt{5},$$

$$L = \int_0^{2\pi} \sqrt{x'(t)^2 + y'(t)^2 + z'(t)^2}\ dt$$

$$= \int_0^{2\pi} \sqrt{4\sin^2 t + 4\cos^2 t + 5}\ dt = \int_0^{2\pi} \sqrt{4+5}\ dt = 3 \cdot 2\pi = 6\pi.$$

■■■**2-68**

A particle is moving along the curve described by the parametric equations $x = 5t$, $y = 2t^3$, and $z = \frac{3}{5}t^5$. Determine the velocity and acceleration vectors as well as the speed of the particle when t = 3. Also, draw a partial sketch of the curve when t = 3 showing the representations of both velocity and acceleration at that point.

****

A vector equation of the given curve can be written as:

$$R(t) = 5t\,\hat{i} + 2t^3\,\hat{j} + \frac{3}{5}t^5\,\hat{k} \quad \dots (1)$$

Hence, we have for velocity:

$$V(t) = D_t\,R(t) = 5\hat{i} + 6t^2\hat{j} + 3t^4\hat{k} \quad \dots (2)$$

Also, for acceleration, we write:

$$A(t) = D_t\,V(t) = 12t\,\hat{j} + 12t^3\,\hat{k} \quad \dots (3)$$

Now, from Eq.(2) we can also write:

$$|V(t)| = \sqrt{(5)^2 + (6t^2)^2 + (3t^4)^2} \quad \dots (4)$$

Using Eqs. (2)–(4), we evaluate, for t = 3

from Eq.(2), $V = 5\hat{i} + 54\hat{j} + 243\hat{k}$ <u>Ans.</u>

from Eq.(3), $A = 36\hat{j} + 324\hat{k}$ <u>Ans.</u>

from Eq.(4),

$$|V(3)| = \sqrt{(5)^2 + \left[6(3)^2\right]^2 + \left[3(3)^4\right]^2}$$
$$= 248.97 \qquad \underline{Ans.}$$

The coordinates x, y, z for t = 3 is calculated as:

$$x = 5(3) = 15$$
$$y = 2(3)^2 = 18$$
$$z = \frac{3}{5}(3)^5 = 145.8$$

The required sketch for the parametric equation is shown alongside.

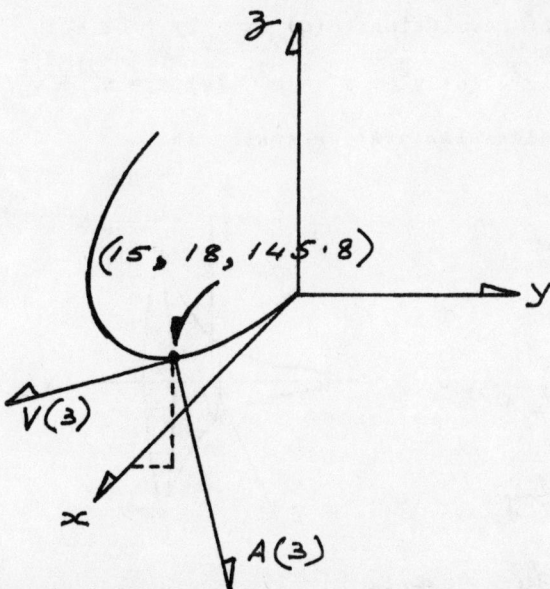

CYLINDERS AND SURFACES OF REVOLUTION

2-69 ▬▬▬▬▬▬▬▬▬▬▬▬▬▬▬▬▬▬▬▬▬▬▬▬▬▬▬▬▬▬▬

Which of the following graphs as a cylinder in space? (a) $x^2 + y^2 + z^2 = 1$

(b) $x = ye^z$ (c) $x = \sin z$ (d) $z = xy$ (e) $z = x^2 + y^2$.

****

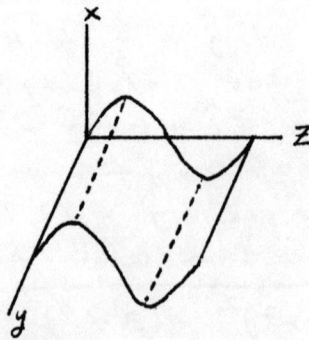

The cylinder is $x = \sin z$, recognized as such because of involving only z of the variables x, y, z.

2-70 ▬▬▬▬▬▬▬▬▬▬▬▬▬▬▬▬▬▬▬▬▬▬▬▬▬▬▬▬▬▬▬

Which of the following is a surface of revolution? (a) $3x - 2y + 5z = 1$

(b) $x^2/4 + y^2/16 + z^2/9 = 1$ (c) $y = z^2$ (d) $y^2 + z^2 = e^{2x}$ (e) $z = x^2 + y^2/3$.

****

$y^2 + z^2 = e^{2x}$ is obtained

from $y = e^x$ by replacing

y with $\sqrt{y^2 + z^2}$, so $y^2 + z^2 = e^{2x}$

is the surface obtained by

revolving $y = e^x$ about the x axis.

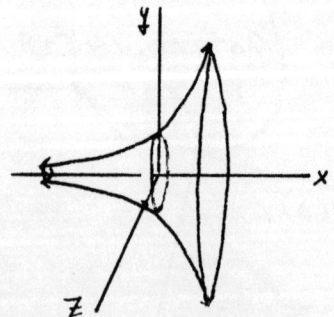

MISCELLANEOUS PROBLEMS

■■ **2-71**

Let L be the line given by x = 2-t, y = 1+t, and z = 1+2t.
L intersects the plane 2x + y - z = 1 at the point P=(1,2,3).

(a) Find the angle L makes with the plane to the nearest
degree.

(b) Find parametric equations for the line through P which
lies in the plane and is perpendicular to L.

****

(a) L is parallel to $\vec{v} = \langle -1, 1, 2 \rangle$

A normal to the plane is $\vec{n} = \langle 2, 1, -1 \rangle$

$\vec{v} \cdot \vec{n} = -3$, $\| \vec{v} \| = \| \vec{n} \| = \sqrt{6}$

∴ angle between $\vec{v}, \vec{n}$ = $\arccos \left(\frac{-3}{6} \right) = \frac{2\pi}{3}$

$\vec{v}$ makes angles of $60°$, $120°$ with $\pm \vec{n}$

∴ $\vec{v}$ makes a $30°$ angle with the plane.

(b) Desired line is through $P = (1, 2, 3)$ and

in direction of $\vec{n} \times \vec{v} = \langle 3, -3, 3 \rangle$.

Equations: x = 1+3t, y = 2-3t, z = 3+3t.

2-72 ■■■

Find the equation of the plane containing the lines x = 4-4t, y = 3-t, z = 1+5t and x = 4-t, y = 3+2t, z = 1.

****

WRITING IN VECTOR NOTATION.

$L_1: \underline{x} = (4,3,1) + t(-4,-1,5) = \underline{x}_1 + t\underline{v}_1$

$L_2: \underline{x} = (4,3,1) + t(-1,2,0) = \underline{x}_2 + t\underline{v}_2$

NOTE THAT $X_1 = X_2 = (4,3,1)$ so $(4,3,1)$ is A POINT ON THE PLANE

$\underline{N}$ = NORMAL TO PLANE IS ORTHOGONAL TO $\underline{v}_1$ AND TO V_2 so.

$$\underline{N} = \underline{V}_1 \times \underline{V}_2 = \begin{pmatrix} i & J & k \\ -4 & -1 & 5 \\ -1 & -2 & 0 \end{pmatrix} = i(0-(-10)) +$$

$$-J(0-(-5)) + \underline{k}(8-1)$$

$$\underline{N} = (10,-5,7)$$

$$\underline{N} \cdot (\underline{x} - \underline{x}_0) = (10,-5,7) \cdot (x-4, y-3, z-1).$$

$N \cdot (x - X_s) = 0$ is EQN. of PLANE

so

$$10x - 5y + 7z - 32 = 0$$

■■■2-73

The position of a particle at time t is given by

$$\vec{r}(t) = t\cos t\,\vec{i} + t\sin t\,\vec{j} + t\vec{k}$$

a) What is the speed of the particle at t = 0?

b) What is the curvature at t = 0?

****

a) Speed $= ds/dt = \|\vec{v}\| = \|d\vec{r}/dt\|$

Now $d\vec{r}/dt = (-t\sin t + \cos t)\vec{i} + (t\cos t + \sin t)\vec{j} + \vec{k}$

So $\|d\vec{r}/dt\| = \sqrt{(-t\sin t + \cos t)^2 + (t\cos t + \sin t)^2 + 1^2}$

$$= \sqrt{t^2 + 2}$$

When $t = 0$ the speed of the particle is $\underline{\sqrt{2}}$.

b) Recall that curvature (K) is given by

$$K = \|\vec{v} \times \vec{a}\| / \|\vec{v}\|^3$$

Now $\vec{v}(0) = d\vec{r}/dt\big|_{t=0} = \vec{i} + \vec{k}$

And $\vec{a} = d^2\vec{r}/dt^2 = (-t\cos t - 2\sin t)\vec{i} + (-t\sin t + 2\cos t)\vec{j}$

So $\vec{a}(0) = 2\vec{j}$

Thus $\|\vec{v} \times \vec{a}\| = \|(\vec{i} + \vec{k}) \times 2\vec{j}\|$

$$= \|-2\vec{i} + 2\vec{k}\| = 2\sqrt{2}$$

And since $\|\vec{v}(0)\| = \sqrt{2}$

we have $K = \dfrac{2\sqrt{2}}{(\sqrt{2})^3} = \underline{1.}$

2-74 ▪▪▪▪▪▪▪▪▪▪▪▪▪▪▪▪▪▪▪▪▪▪▪▪▪▪▪▪▪▪▪▪▪▪▪▪▪▪▪

Given the points in space :
  A = (2,3,1);   B = (4,-1,5); O = (0,0,0).
Find the distance from O to the line through A and B.

****

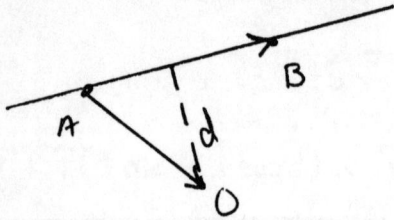

d is the length of the component of $\vec{AO}$ in the direction $\perp$ to $\vec{AB}$, thus

$$d = \frac{|\vec{AO} \times \vec{AB}|}{|\vec{AB}|} \qquad \vec{AO} = \langle -2,-3,-1 \rangle$$

$$\vec{AB} = \langle 2,-4,4 \rangle$$

$$\therefore \ |\vec{AO} \times \vec{AB}| = \left| \begin{vmatrix} i & j & k \\ -2 & -3 & -1 \\ 2 & -4 & 4 \end{vmatrix} \right| = \left| -16\vec{i} + 6\vec{k} + 14\vec{j} \right|$$

$$= \sqrt{488}$$

and $|\vec{AB}| = \sqrt{36} = 6$

thus $d = \dfrac{\sqrt{488}}{6} = \boxed{\dfrac{\sqrt{122}}{3}}$

■■■ **2-75**

Find the volume of the parallelepiped spanned by the vectors $2\vec{i} + 3\vec{j} + 5\vec{k}$, $3\vec{i} - \vec{j} + 4\vec{k}$, and $-\vec{i} - 2\vec{j} + 3\vec{k}$.

****

Recall that if a parallelepiped is spanned by the vectors $\vec{x}$, $\vec{y}$ and $\vec{w}$, then the volume of the parallelepiped is given by the triple product

$$\left| \vec{x} \cdot (\vec{y} \times \vec{w}) \right|.$$

Here, $\vec{x} = 2\vec{i} + 3\vec{j} + 5\vec{k}$,

$$\vec{y} = 3\vec{i} - \vec{j} + 4\vec{k},$$

$$\vec{w} = -\vec{i} - 2\vec{j} + 3\vec{k}$$

$$\vec{y} \times \vec{w} = \begin{vmatrix} \vec{i} & \vec{j} & \vec{k} \\ 3 & -1 & 4 \\ -1 & -2 & 3 \end{vmatrix} = 5\vec{i} - 13\vec{j} - 7\vec{k}$$

$$\left| \vec{x} \cdot (\vec{y} \times \vec{w}) \right| = \left| 2(5) + (-1)(-13) + (4)(-7) \right|$$

$$= \left| -5 \right| = 5 \text{ cubic units}$$

$$= \text{volume}$$

2-76 ■■

Suppose that a dove is flying so that its **acceleration** vector at time t
is given by a(t)= (t^4 , t^3 , t^2), its initial velocity and its initial
displacement are given by v(0)= (1, 2, 3) and s(0)= (7, 6, 5) respectively.
Find the position vector s(t) at time t.

$$\ast\ast\ast\ast\ast\ast\ast\ast\ast\ast\ast\ast\ast\ast\ast\ast\ast\ast\ast\ast\ast\ast\ast\ast\ast\ast\ast\ast\ast\ast\ast\ast\ast\ast\ast\ast\ast\ast\ast\ast\ast\ast\ast\ast\ast\ast$$

$$\text{Since} \quad a(t) = \left(t^4, t^3, t^2 \right)$$

$$V(t) = \left(\frac{t^5}{5} + C_1, \frac{t^4}{4} + C_2, \frac{t^3}{3} + C_3 \right); \text{hence}$$

$$V(0) = (1, 2, 3) \text{ implies } (C_1, C_2, C_3) = (1, 2, 3).$$

$$\text{So} \quad V(t) = \left(\frac{t^5}{5} + 1, \frac{t^4}{4} + 2, \frac{t^3}{3} + 3 \right). \text{ Therefore}$$

$$S(t) = \left(\frac{t^6}{30} + t + k_1, \frac{t^5}{20} + 2t + k_2, \frac{t^4}{12} + 3t + k_3 \right). \text{ So}$$

$$S(0) = (7, 6, 5) \text{ implies } (k_1, k_2, k_3) = (7, 6, 5);$$

$$\text{hence,} \quad S(t) = \left(\frac{t^6}{30} + t + 7, \frac{t^5}{20} + 2t + 6, \frac{t^4}{12} + 3t + 5 \right).$$

2-77 ■■

If a force $\vec{F} = \vec{i} + 2\vec{j} + 5\vec{k}$ moves an object from P = (1,−2,4) to
Q = (3,2,5) along segment $\overrightarrow{PQ}$, then the work done is   (a) 15   (b) $\sqrt{17}$
(c) 32   (d) $\sqrt{21}$ (e) 49.

$$\ast\ast\ast\ast\ast\ast\ast\ast\ast\ast\ast\ast\ast\ast\ast\ast\ast\ast\ast\ast\ast\ast\ast\ast\ast\ast\ast\ast\ast\ast\ast\ast\ast\ast\ast\ast\ast\ast\ast\ast\ast\ast\ast\ast\ast\ast$$

$$\overrightarrow{PQ} = \overrightarrow{OQ} - \overrightarrow{OP} = \langle 3,2,5 \rangle - \langle 1,-2,4 \rangle = \langle 2,4,1 \rangle,$$

$$\text{work} = \vec{F} \cdot \overrightarrow{PQ} = \langle 1,2,5 \rangle \cdot \langle 2,4,1 \rangle = 2+8+5 = 15.$$

2-78

Find an equation of the plane with nonzero intercepts (a,0,0), (0,b,0), and (0,0,c). Also show that the resulting equation can be written in the form :

$$\frac{x}{a} + \frac{y}{b} + \frac{z}{c} = 1$$

****

The equation of the plane with nonzero intercepts can be written as

$$A(x-x_1) + B(y-y_1) + C(z-z_1) = 0 \quad \ldots (1)$$

OR
$$Ax + By + Cz = Ax_1 + By_1 + Cz_1 = D$$

where D is a constant

OR, more generally as :

$$Ax + By + Cz = D \quad \ldots\ldots (2)$$

which is known as the general equation of a plane.

By substituting the three given values, we get :

$$A(a) = D \quad ; \quad A = \frac{D}{a},$$
$$B(b) = D \quad ; \quad B = \frac{D}{b},$$
and
$$C(c) = D \quad ; \quad C = \frac{D}{c}.$$

By rewriting Eq. (1), we therefore have

$$\frac{D}{a}(x) + \frac{D}{b}(y) + \frac{D}{c}(z) = D$$

OR
$$\frac{x}{a} + \frac{y}{b} + \frac{z}{c} = 1.$$

2-79

Show that the line

    x = 1 + 2t
    y = -1 + 3t and the plane x - 2y + z = 6
    z = 2 + 4t

are parallel. Find the distance between them.

****

The normal to the plane is $\langle 1, -2, 1 \rangle$ and the line is parallel to $\langle 2, 3, 4 \rangle$. Since $\langle 1, -2, 1 \rangle \cdot \langle 2, 3, 4 \rangle = 0$, the line and the plane are parallel. Taking any point on the plane, like $(6, 0, 0)$ and subtracting a point on the line, like $(1, -1, 2)$, we obtain $\langle 5, 1, -2 \rangle$, a vector joining the line to the plane. Projecting onto $\langle 1, -2, 1 \rangle$, the distance is

$$\langle 5, 1, -2 \rangle \cdot \frac{1}{\sqrt{6}} \langle 1, -2, 1 \rangle = \frac{1}{\sqrt{6}}.$$

██**2-80**

Given points A(1,2,3,), B(4,0,1), and C(2,2,7) find:

a. the distance from A to B.
b. symmetric equations of the line on A and B.
c. the equation of the plane on A, B, and C.
d. the area of the triangle with vertices A, B, and C.

****

a. $d = \sqrt{(4-1)^2 + (0-2)^2 + (1-3)^2} = \sqrt{9+4+4} = \sqrt{17}$

b. The vector in the direction of this line is $(4-1)i + (0-2)j + (1-3)k$ or $3i - 2j - 2k$. There are two reasonable answers to this question depending on which point is used as a reference.

If A: $\frac{x-1}{3} = \frac{y-2}{-2} = \frac{z-3}{-2}$. If B: $\frac{x-4}{3} = \frac{y}{-2} = \frac{z-1}{-2}$

c. We need the components of a vector orthogonal to the plane. This is found by taking the cross-product of two vectors in the plane. From part b, one of them is $3i - 2j - 2k$. Another could be the vector from A to C, which is

$(2-1)i + (2-2)j + (7-3)k$ or $i + 4k$.

So $(i+4k) \times (3i - 2j - 2k) = [0(-2) - 4(-2)]i + [4(3) - 1(-2)]j$
$$+ [1(-2) - 0(3)]k$$
$$= 8i + 14j - 2k$$

We use these components with the coordinates of A to get the equation of the plane:
$$8(x-1) + 14(y-2) - 2(z-3) = 0$$
$$8x + 14y - 2z - 30 = 0$$
So, best answer is $4x + 7y - z - 15 = 0$

d. Since the norm (magnitude) of the cross-product of two vectors gives the area of the parallelogram determined by them, the area of the triangle is one half of this. Using the result from part c

area of $\triangle ABC = \frac{1}{2} \| 8i + 14j - 2k \| = \frac{1}{2}\sqrt{8^2 + 14^2 + (-2)^2}$
$$= \frac{1}{2}\sqrt{264} = \frac{1}{2} \cdot 2\sqrt{66} = \sqrt{66} \text{ square units}$$

2-81 ■■

Find the volume of the tetrahedron with vertices (0, 0, 0), (1, 2, 0), (-1, 1, 1), and (1, -2, 2).

****

The vectors on three edges are

$$\vec{V_1} = \langle 1, 2, 0 \rangle, \quad \vec{V_2} = \langle -1, 1, 1 \rangle, \text{ and } \vec{V_3} = \langle 1, -2, 2 \rangle.$$

The volume is $\frac{1}{6} \vec{V_1} \cdot (\vec{V_2} \times \vec{V_3})$, the scalar triple product, which is the determinant of

$$\begin{bmatrix} 1 & 2 & 0 \\ -1 & 1 & 1 \\ 1 & -2 & 2 \end{bmatrix}, \text{ which is}$$

$$1 \cdot 1 \cdot 2 + 2 \cdot 1 \cdot 1 + 0 \cdot (-1) \cdot (-2) - 1 \cdot 1 \cdot 0 - 1 \cdot 1 \cdot (-2)$$

$$- (-1) \cdot 2 \cdot 2 = 10, \text{ so the volume is}$$

$$\frac{10}{6} = 1\frac{2}{3}.$$

2-82

Find the distance between the two skew lines:

$$L_1: \frac{x-1}{2} = \frac{y-3}{5} = \frac{z-1}{3} \quad \text{and} \quad L_2: \frac{x-2}{4} = \frac{y+1}{2} = \frac{z+2}{3}$$

****

The distance desired is along a vector perpendicular to both lines. $\vec{A} = 2\vec{i} + 5\vec{j} + 3\vec{k}$ is parallel to L_1, $\vec{B} = 4\vec{i} + 2\vec{j} + 3\vec{k}$ is parallel to L_2, and $\vec{A} \times \vec{B}$ is perpendicular to both L_1 and L_2.

$$\vec{C} = \vec{A} \times \vec{B} = \begin{vmatrix} \vec{i} & \vec{j} & \vec{k} \\ 2 & 5 & 3 \\ 4 & 2 & 3 \end{vmatrix} = \vec{i} \begin{vmatrix} 5 & 3 \\ 2 & 3 \end{vmatrix} - \vec{j} \begin{vmatrix} 2 & 3 \\ 4 & 3 \end{vmatrix} + \vec{k} \begin{vmatrix} 2 & 5 \\ 4 & 2 \end{vmatrix}$$

$$= \vec{i}(15-6) - \vec{j}(6-12) + \vec{k}(4-20) = 9\vec{i} + 6\vec{j} - 16\vec{k}$$

A vector that originates on L_1 and terminates on L_2 is

$$\vec{D} = (2-1)\vec{i} + (-1-3)\vec{j} + (-2-1)\vec{k} = \vec{i} - 4\vec{j} - 3\vec{k}$$

The distance between the lines is the scalar projection of $\vec{D}$ onto $\vec{C}$ (or the component of $\vec{D}$ in the direction of $\vec{C}$)

$$\text{distance} = \frac{\vec{D} \cdot \vec{C}}{|\vec{C}|} = \frac{1 \cdot 9 + -4 \cdot 6 + -3 \cdot -16}{\sqrt{9^2 + 6^2 + (-16)^2}} = \frac{9 - 24 + 48}{\sqrt{373}}$$

$$= \frac{33}{\sqrt{373}} = \frac{33\sqrt{373}}{373}$$

2-83 ■■

Find the volume of the parallelopiped given P = (1,-3,2), Q = (3,-1,3), R = (2,1,-4), S = (-1,2,1).

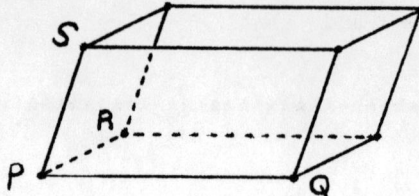

****

$\overrightarrow{PQ} = \langle 2,2,1 \rangle$, $\overrightarrow{PR} = \langle 1,4,-6 \rangle$, $\overrightarrow{PS} = \langle -2,5,-1 \rangle$. The

Volume is given by the triple scalar product,

$(\overrightarrow{PQ} \times \overrightarrow{PR}) \cdot \overrightarrow{PS}$. $\overrightarrow{PQ} \times \overrightarrow{PR} = \begin{vmatrix} \vec{i} & \vec{j} & \vec{k} \\ 2 & 2 & 1 \\ 1 & 4 & -6 \end{vmatrix} = \langle -16,13,6 \rangle$, so

$V = \langle -16,13,6 \rangle \cdot \langle -2,5,-1 \rangle = 32+65-6 = 91$

3

FUNCTIONS OF SEVERAL VARIABLES AND PARTIAL DERIVATIVES

FUNCTIONS OF SEVERAL VARIABLES

∎∎**3-1**

State in set builder notation and sketch the domain of $f(x,y,z) = \sqrt{xy} \sin z$.

****

Domain of $f = \{ (x,y,z) \mid xy \geq 0, \; x,y,z \in \text{Reals} \}$.

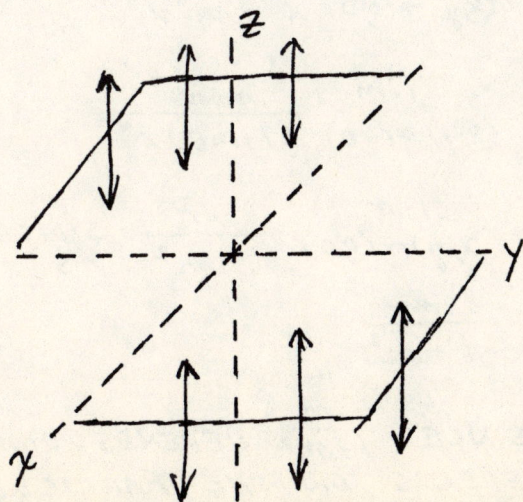

The I, III, V, and VII octants are the domain of f.

105

3-2 ■■■

Let $f(x,y) = \dfrac{xy}{x^2 + y^2}$ and let C_m be the "curve" with equation $y = mx$ where m is a constant. The value of the limit of $f(x,y)$ as (x,y) approaches $(0,0)$ along C_m is (a) 0 (b) 1/2 (c) 1 (d) $m/(1 + m^2)$ (e) $m^2/(1 + m^2)$.

****

$$\lim_{\substack{(x,y)\to(0,0)\\ \text{on } y=mx}} \frac{xy}{x^2+y^2} = \lim_{x\to 0} \frac{x(mx)}{x^2+(mx)^2} = \lim_{x\to 0} \frac{mx^2}{(1+m^2)x^2} = \frac{m}{1+m^2}$$

LIMITS OF FUNCTIONS OF SEVERAL VARIABLES

3-3 ■■■

Find the limit as $(x,y)\to(0,0)$ of $\dfrac{x^2 y}{x^4+y^2}$ or show that no limit exists.

****

LET $y = mx^2$. THEN

$$\lim_{(x,y)\to(0,0)} \frac{x^2 y}{x^4+y^2} = \lim_{(x,y)\to(0,0)} \frac{x^2 \, mx^2}{x^4+m^2 x^4}$$

$$= \lim_{(x,y)\to(0,0)} \frac{mx^4}{(1+m^2)x^4}$$

$$= \lim_{(x,y)\to(0,0)} \frac{m}{1+m^2} \quad \text{SINCE } x\neq 0 \text{ AS } x\to 0$$

$$= \frac{m}{1+m^2}$$

NO LIMIT EXISTS SINCE $\frac{m}{1+m^2}$ DEPENDS ON m. THAT IS, $\frac{m}{1+m^2}$ DEPENDS ON THE PARTICULAR CURVE ALONG WHICH (x,y) APPROACHES $(0,0)$.

3-4

Let f(x,y) be defined by:

$$f(x,y) = \begin{cases} \dfrac{x^4 + y^4}{(x^2 + y^2)^2}, & (x,y) \neq (0,0) \\[2mm] 0, & (x,y) = (0,0) \end{cases}$$

Does this function have a limit at the origin? If so, prove it. If not, demonstrate why not.

****

Approaching the origin along $y = 0$,

$$\lim_{(x,y) \to (0,0)} \frac{x^4 + y^4}{(x^2 + y^2)^2} = \lim_{x \to 0} \frac{x^4}{x^4} = 1$$

Approaching the origin along $y = x$,

$$\lim_{(x,y) \to (0,0)} \frac{x^4 + y^4}{(x^2 + y^2)^2} = \lim_{x \to 0} \frac{x^4 + x^4}{(x^2 + x^2)^2} = \lim_{x \to 0} \frac{2x^4}{4x^4}$$

$$= \frac{1}{2}$$

$\therefore$ This function does <u>not</u> have a limit at the origin.

3-5 ■■■

Prove that the following limit does not exist

$$\lim_{(x,y)\to(0,0)} f(x,y) \quad \text{where } f(x,y) = \frac{xy}{x^2 + y^4}$$

****

f is defined everywhere in R^2 except at $(0,0)$.

Let $S_1 = \{(0,y)\}$, then $(0,0) \in S_1$ and $\lim\limits_{\substack{(x,y)\to(0,0)\\ P \in S_1}} f(x,y) =$

$$\lim_{y\to 0} \frac{0 \cdot y}{0^2 + y^4} = \lim_{y\to 0} \frac{0}{y^4} = \lim_{y\to 0} 0 = 0$$

Let $S_2 = \{(x,x)\}$, then $(0,0) \in S_2$, and $\lim\limits_{\substack{(x,y)\to(0,0)\\ P \in S_2}} f(x,y) =$

$$\lim_{x\to 0} \frac{x \cdot x}{x^2 + x^4} = \lim_{x\to 0} \frac{1}{1 + x^2} = 1$$

Because $\lim\limits_{\substack{(x,y)\to(0,0)\\ P \in S_1}} f(x,y) \neq \lim\limits_{\substack{(x,y)\to(0,0)\\ P \in S_2}} f(x,y)$,

the limit does not exist.

■■ **3-6**

Show that the limit

$$\lim_{(x,y) \to (0,0)} \frac{2x^2 - y^2}{x^2 + 2y^2}$$

does not exist.

****

Approaching (0,0) along the x-axis we have y = 0

so $\lim_{(x,y) \to (0,0)} \dfrac{2x^2 - y^2}{x^2 + 2y^2} = \lim_{(x,y) \to (0,0)} \dfrac{2x^2}{x^2} = 2$

Approaching (0,0) along the y-axis we have x = 0

so $\lim_{(x,y) \to (0,0)} \dfrac{2x^2 - y^2}{x^2 + 2y^2} = \lim_{(x,y) \to (0,0)} \dfrac{-y^2}{2y^2} = -\dfrac{1}{2}$

Since these limits are different,

$$\lim_{(x,y) \to (0,0)} \frac{2x^2 - y^2}{x^2 + 2y^2}$$

does not exist.

3-7 ■■■

If $f(x,y) = \dfrac{xy}{x^2 + y^2}$ then $\lim\limits_{(x,y)\to(0,0)} f(x,y)$ (a) exists (b) does not exist

(c) is equal to 0   (d) is equal to 1/2   (e) is equal to 1.

****

$\lim\limits_{(x,y)\to(0,0)} f(x,y)$ does not exist since different limit values

are obtained on different paths approaching $(0,0)$.

For example, $\lim\limits_{\substack{(x,y)\to(0,0)\\ \text{on } y=x}} \dfrac{xy}{x^2+y^2} = \lim\limits_{x\to 0} \dfrac{x^2}{2x^2} = \dfrac{1}{2}$, while

$\lim\limits_{\substack{(x,y)\to(0,0)\\ \text{on } y=2x}} \dfrac{xy}{x^2+y^2} = \lim\limits_{x\to 0} \dfrac{2x^2}{5x^2} = \dfrac{2}{5}$.

CONTINUITY OF FUNCTIONS
OF SEVERAL VARIABLES

3-8 ■■■

Let $f(x,y) = 3x + 7y$. For any $\epsilon > 0$, $0 < \sqrt{(x-2)^2 + (y-3)^2} < \delta$ implies $|f(x,y) - f(2,3)| < \epsilon$ provided (a) $\delta = \epsilon$ (b) $\delta = 2\epsilon$ (c) $\delta = \epsilon/10$
(d) $\delta = \epsilon/2$ (e) $\delta = \epsilon/3$.

****

$0 < \sqrt{(x-2)^2 + (y-3)^2} < \delta$ implies $|f(x,y) - f(2,3)| =$

$|3x + 7y - 27| = |3(x-2) + 7(y-3)| \leq |3(x-2)| + |7(y-3)| =$

$3|x-2| + 7|y-3| \leq 3\sqrt{(x-2)^2+(y-3)^2} + 7\sqrt{(x-2)^2+(y-3)^2} <$

$3\delta + 7\delta = 10\delta = \epsilon$ provided $\delta = \epsilon/10$.

━━━━━━━━━━━━━━━━━━━━━━━━━━━━━━━━━━━━ 3-9

Determine if the following function is everywhere continuous, and if not, locate the point(s) of discontinuity:

$$f(x,y) = \begin{cases} \dfrac{x^2 + yx}{x^2 + y^2} & \text{if } (x\ y) \neq (0,0) \\[2mm] 1 & \text{if } (x,y) = (0,0) \end{cases}$$

****

At all points $(x,y) \neq (0,0)$, $f(x,y)$ is a rational function, and is continuous where its denominator does not equal zero (which happens only at $(0,0)$), so the only point which might be a discontinuous point is $(0,0)$. We must determine if the limit exists at $(0,0)$, so we inspect limits as we approach $(0,0)$ along various curves.

$y = x$: $\displaystyle \lim_{\substack{(x,y) \to (0,0) \\ y=x}} \frac{x^2 + yx}{x^2+y^2} = \lim_{x \to 0} \frac{x^2+x^2}{x^2+x^2} = \lim_{x \to 0} 1 = 1$

$x = 0$: $\displaystyle \lim_{\substack{(x,y) \to (0,0) \\ x=0}} \frac{x^2+yx}{x^2+y^2} = \lim_{y \to 0} \frac{0+0}{0+y^2} = \lim_{y \to 0} 0 = 0$

Since we found two curves which give us different values, we conclude that the limit does not exist at $(0,0)$, and thus $(0,0)$ is a discontinuous point.

3-10 ▪▪

If $f(x,y) = \dfrac{y}{\sqrt{x^2 + y^2 - 16}}$, briefly discuss the continuity of f. Also,

draw a sketch to show as a shaded region in $\mathcal{R}$ the region where f is continuous.

****

We note from the given equation that its domain is the set of all points in the exterior region of the circle described by $x^2 + y^2 = 16$ and all points on the x-axis such that $-4 < x < 4$.

Also, the given function can be considered as a quotient of functions g & h such that : $g(x,y) = y$

and $h(x,y) = \sqrt{x^2 + y^2 - 16}$

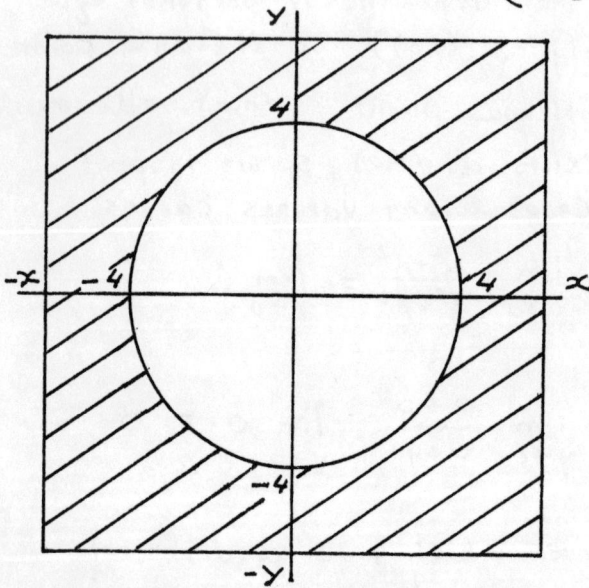

We note from the above and the sketch shown alongside that $g(x,y)$ is a polynomial, which makes it continuous at all points, and h is continuous at any and all points in $\mathcal{R}^2$ such that $x^2 + y^2 > 16$.

Hence we conclude that f is continuous at all points outside the region bounded by the circle ($x^2 + y^2 = 16$).

By considering the points on the x-axis such that $-4 < x < 4$ (i.e. the points $(a,0)$ where $-4 < a < 4$), we note that

$$\lim_{(x,y) \to (a,0)} f(x,y) = \lim_{y \to 0} \frac{y}{\sqrt{a^2 + y^2 - 16}}$$

However, the above does not exist because

$$\frac{y}{\sqrt{a^2+y^2-16}}$$ is undefined if $|y| \leq \sqrt{16-a^2}$.

Hence, we conclude that f is discontinuous at those points on the x-axis such that $-4 < x < 4$.

-- **3-11**

Determine whether the following function is continuous at $(0,0)$.

$$f(x,y) = \begin{cases} \dfrac{3x^2 - 2y^2}{x^2 + y^2} & \text{if } (x,y) \neq (0,0) \\ 0 & \text{if } (x,y) = (0,0) \end{cases}$$

****

TRY THE PATH $y = mx$.

$$\lim_{(x,y)\to(0,0)} f(x,y) = \lim_{x\to 0} \frac{3x^2 - 2m^2x^2}{x^2 + m^2x^2} = \lim_{x\to 0} \frac{3-2m^2}{1+m^2}$$

$$= \frac{3-2m^2}{1+m^2}$$

THIS DEPENDS UPON m. $\therefore \lim_{(x,y)\to(0,0)} f(x,y)$ DOES NOT EXIST AND, HENCE, THE FUNCTION IS DISCONTINUOUS AT $(0,0)$.

GRAPHING FUNCTIONS OF SEVERAL VARIABLES

3-12 ■■■■■■■■■■■■■■■■■■■■■■■■■■■■■■■■■■■■■■

The temperature at a point (x,y) of a flat metal plate is

$$T(x,y) = 9x^2 + 16y^2$$

where T(x,y) is in degrees. Draw the isothermals for T(x,y)=0, 9,16, and 144 degrees.

****

For $T(x,y)=0$, $0 = 9x^2 + 16y^2$ which is a point-ellipse

" $T(x,y)=9$, $9 = 9x^2 + 16y^2$ or $1 = x^2 + \dfrac{y^2}{9/16}$

an ellipse with $a = 1$ and $b = 3/4$

" $T(x,y)=16$, $16 = 9x^2 + 16y^2$ or $1 = \dfrac{x^2}{\frac{16}{9}} + y^2$

an ellipse with $a = 4/3$ and $b = 1$

" $T(x,y) = 144$, $144 = 9x^2 + 16y^2$ or $1 = \dfrac{x^2}{16} + \dfrac{y^2}{9}$

an ellipse with $a = 4$ and $b = 3$

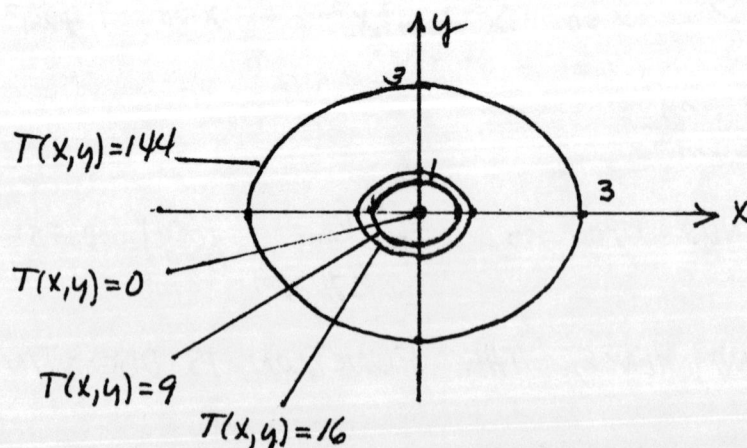

■■■■■■■■■■■■■■■■■■■■■■■■■■■■■■■■■■ 3-13

For the function $z = \sqrt{x^2 + y^2 - 1}$

a) Sketch the domain of the function.

b) Sketch the level curves $z = k$ for $k = 0, 1, 2,$ and 3.

c) Sketch the portion of the surface in the first octant.

****

a) Since $z = \sqrt{x^2 + y^2 - 1}$
we must have
$$x^2 + y^2 - 1 \geq 0.$$
This is the region on
the unit circle and
outside of it.

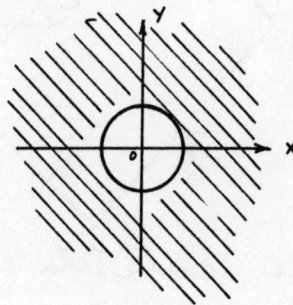

b) Note that when $z = k$
we have $x^2 + y^2 = k^2 + 1$.
So for each value of
k, the level curve is
a circle, centered at the
origin with radius $\sqrt{k^2 + 1}$.

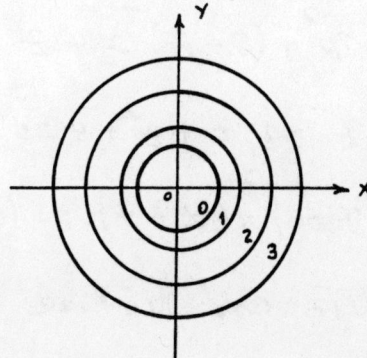

c)   TRACES
    xy-plane: circle, radius 1
    ⎧ xz-plane: $z = \sqrt{x^2 - 1}$
    ⎨
    ⎩ yz-plane: $z = \sqrt{y^2 - 1}$

    └ These are hyperbolas.

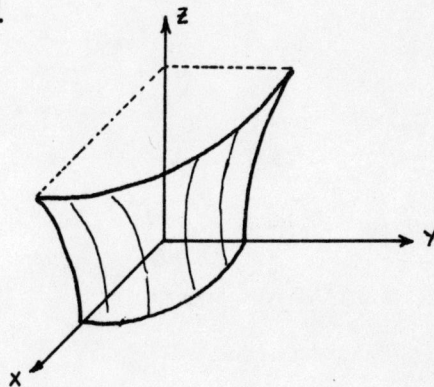

3-14 ■■■

Consider the surfaces $x^2 + y^2 = 4z$ and $z = 2 - y$. Find the projection of the intersection of these surfaces on the xz coordinate plane.

****

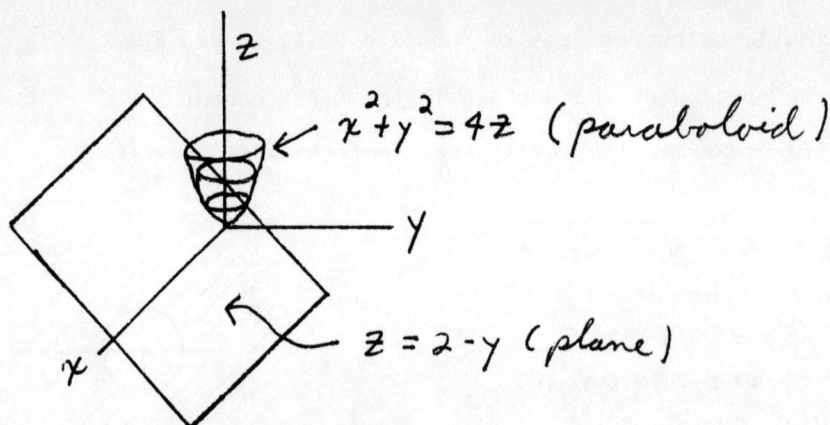

We substitute $y = 2 - z$ into $x^2 + y^2 = 4z$ $\Rightarrow$

$x^2 + (2-z)^2 = 4z$ $\Rightarrow$ $x^2 + 4 - 4z + z^2 = 4z$ $\Rightarrow$

$x^2 - 8z + z^2 + 4 = 0$ $\Rightarrow$ $x^2 + z^2 - 8z + 16 = 12$ $\Rightarrow$

$(x-0)^2 + (z-4)^2 = (\sqrt{12})^2$ a circle of radius

$\sqrt{12}$ and whose center is located at $(0,4)$.

This circle is the desired projection.

■■■ **3-15**

Suppose the point (2,3) is on a curve C which is a level curve of the surface $z = x^2 + 2y$. Can it be concluded that the point (4,-3) is also on C? (a) yes (b) no.

****

Given that C is a level curve of $z = x^2 + 2y$ means

C has equation $x^2 + 2y = K$ for some constant K.

(2,3) on C gives K = 10, so C has equation $x^2 + 2y = 10$

which <u>does</u> contain the point (4, -3).

■■■ **3-16**

Sketch $\phi = \frac{\pi}{6}$ and $\phi = \frac{\pi}{2}$ in three dimensional space.

****

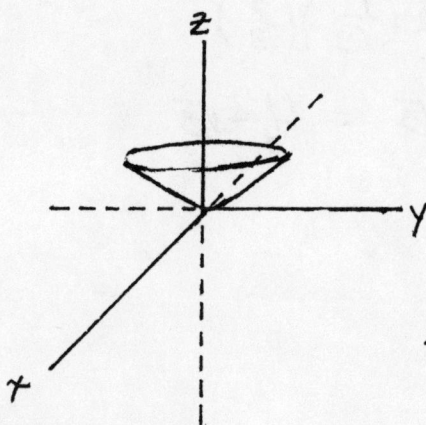

$\phi = \frac{\pi}{6}$ *a cone.* $\phi = \frac{\pi}{2}$ *the xy plane.*

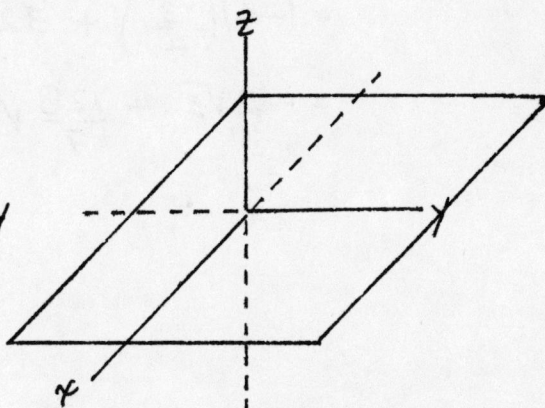

PARTIAL DERIVATIVES

3-17 ■■

If $f(r,\theta) = r\sin(\theta) + r^2\tan^2(\theta)$, find the partial derivative of f with respect to r and also with respect to θ at the point (-4, $\pi/6$).

****

$$\frac{\partial f}{\partial r} = \sin(\theta) + 2r\tan^2(\theta)$$

$$\frac{\partial f}{\partial r}\bigg|_{(-4,\pi/6)} = \sin(\pi/6) + 2(-4)\tan^2(\pi/6) = \frac{1}{2} - 8\left(\frac{1}{\sqrt{3}}\right)^2$$

$$= \frac{1}{2} - \frac{8}{3} = -\frac{13}{6}$$

$$\frac{\partial f}{\partial \theta} = r\cos(\theta) + 2r^2\tan(\theta)\sec^2(\theta)$$

$$\frac{\partial f}{\partial \theta}\bigg|_{(-4,\pi/6)} = (-4)\cos(\pi/6) + 2(-4)^2\tan(\pi/6)\sec^2(\pi/6)$$

$$= (-4)\left(\frac{\sqrt{3}}{2}\right) + 32\left(\frac{1}{\sqrt{3}}\right)\left(\frac{4}{3}\right)$$

$$= -2\sqrt{3} + \frac{128}{9}\sqrt{3} = \frac{110}{9}\sqrt{3}$$

━━━ **3-18**

Define f(x,y) by:

$$f(x,y) = \begin{cases} \dfrac{x^2}{x+y}, & (x,y) \neq (0,0) \\[2mm] 0, & (x,y) = (0,0) \end{cases}$$

(a) Find $\dfrac{\partial f}{\partial x}$ and $\dfrac{\partial f}{\partial y}$.

(b) Find the values of the above derivatives at (0,0), if they exist.

****

For $(x,y) \neq (0,0)$,

$$\frac{\partial f}{\partial x} = \frac{2x(x+y) - x^2}{(x+y)^2} = \frac{x^2 + 2xy}{(x+y)^2}$$

$$\frac{\partial f}{\partial y} = \frac{-x^2}{(x+y)^2}$$

$$\frac{\partial f}{\partial x}(0,0) = \lim_{\Delta x \to 0} \frac{f(\Delta x, 0) - f(0,0)}{\Delta x} = \lim_{\Delta x \to 0} \frac{\Delta x - 0}{\Delta x}$$

$$= 1$$

$$\frac{\partial f}{\partial y}(0,0) = \lim_{\Delta y \to 0} \frac{f(0, \Delta y) - f(0,0)}{\Delta y} = \lim_{\Delta y \to 0} \frac{0-0}{\Delta y} = 0$$

━━━ **3-19**

If $z = r^2 \tan^2\theta$, find $\left(\dfrac{\partial z}{\partial x}\right)_r$. (Where (r,θ) are the usual polar coordinates.)

****

$$z = r^2 \frac{r^2 \sin^2\theta}{r^2 \cos^2\theta} = r^2 \frac{y^2}{x^2} = r^2 \frac{(r^2 - x^2)}{x^2} = \frac{r^4}{x^2} - r^2$$

$$\left(\frac{\partial z}{\partial x}\right)_r = \frac{-2r^4}{x^3}$$

3-20 ▬▬▬▬▬▬▬▬▬▬▬▬▬▬▬▬▬▬▬▬▬▬▬▬▬▬

Find both partial derivatives if

$$f(x,y) = \frac{2x - 3y}{3x - 2y}$$

****

$$\frac{\delta f}{\delta x} = \frac{(3x-2y)2 - (2x-3y)(3)}{(3x-2y)^2} \qquad \frac{\delta f}{\delta y} = \frac{(3x-2y)(-3) - (2x-3y)(-2)}{(3x-2y)^2}$$

$$\frac{\delta f}{\delta x} = \frac{5y}{(3x-2y)^2} \qquad\qquad \frac{\delta f}{\delta y} = \frac{-5x}{(3x-2y)^2}$$

3-21 ▬▬▬▬▬▬▬▬▬▬▬▬▬▬▬▬▬▬▬▬▬▬▬▬▬▬

If $w = x^2z + xy^2 - yz^2$, find $\partial w/\partial x$, $\partial w/\partial y$, and $\partial w/\partial z$.

****

Regarding y and z as constants,

$$\frac{\partial w}{\partial x} = 2zx + y^2$$

Regarding x and z as constants,

$$\frac{\partial w}{\partial y} = 2xy - z^2$$

Regarding x and y as constants,

$$\frac{\partial w}{\partial z} = x^2 - 2yz$$

■■ **3-22**

Given $f(x,y) = 2x^3y^2 - 3x^3 + 8y^4$, find f_x and f_y and evaluate each at $(1,2)$

$***$

f_x : hold y constant and differentiate with respect to x.

$$f_x = 2y^2(3x^2) - 9x^2 = 6x^2y^2 - 9x^2$$

$$f_x(1,2) = 6(1^2)(2^2) - 9(1^2) = 15$$

f_y: hold x constant and differentiate with respect to y.

$$f_y = 2x^3(2y) + 32y^3 = 4x^3y + 32y^3$$

$$f_y(1,2) = 4(1^3)(2) + 32(2^3) = 264$$

■■ **3-23**

Let $f(x, y, z) = x^2y^3 - \dfrac{x}{z} + e^x \ln y$.
(a) Find $f_x(x, y, z)$.
(b) Find $f_y(x, y, z)$.
(c) Find $f_z(x, y, z)$.

$***$

(a) $f_x(x, y, z) = 2xy^3 - \dfrac{1}{z} + e^x \ln y$

(b) $f_y(x, y, z) = 3x^2y^2 + e^x\left(\dfrac{1}{y}\right)$

$$= 3x^2y^2 + \dfrac{e^x}{y}$$

(c) $f_z(x, y, z) = \dfrac{x}{z^2}$

3-24 ■■

Let D_i denote partial differentiation with respect to the ith variable.
If $f(x,y) = x^2y^3$ then $D_1f(x,y) + D_2f(x,y)$ is   (a) $xy^2(2y + 3x)$
(b) $2x + 3y^2$ (c) $x^2 + y^3$ (d) $2xy^3$ (e) $3x^2y^2$.

****

Treating y as constant and differentiating with respect to x, $D_1f(x,y) = 2xy^3$. Similarly, $D_2f(x,y) = 3x^2y^2$, so $D_1f + D_2f = 2xy^3 + 3x^2y^2 = xy^2(2y+3x)$.

HIGHER ORDER PARTIAL DERIVATIVES

3-25 ■■■

Given $f(x,y) = x^2\sin(xy)$, find
   a. the partial derivative of f with respect to x,
   b. the partial derivative of f with respect to y,
   c. the second partial derivative of f, first with respect to
       x, then with respect to y.

****

a) $2x\sin(xy) + x^2y\cos(xy)$

b) $x^3\cos(xy)$

c) DIFFERENTIATING THE ANSWER TO PART a) WITH RESPECT TO y,

$2x^2\cos(xy) + x^2\cos(xy) - x^3y\sin(xy)$
$= 3x^2\cos(xy) - x^3y\sin(xy)$

■■■ **3-26**

Let $f(x,y,z) = (xyz)^2$. Find all second order partial derivatives.

****

$$\frac{\delta f}{\delta x} = 2(xyz)yz = 2xy^2z^2 \qquad \frac{\delta f}{\delta y} = 2(xyz)(xz) = 2x^2yz^2$$

$$\frac{\delta f}{\delta z} = 2(xyz)(xy) = 2x^2y^2z$$

$$\frac{\delta^2 f}{\delta x^2} = 2y^2z^2 \qquad\qquad \frac{\delta^2 f}{\delta y^2} = 2x^2z^2$$

$$\frac{\delta^2 f}{\delta z^2} = 2x^2y^2$$

$$\frac{\delta^2 f}{\delta y \, \delta x} = 2xz^2(2y) = 4xyz^2 \qquad \frac{\delta^2 f}{\delta z \, \delta x} = 2xy^2(2z) = 4xy^2z$$

$$\frac{\delta^2 f}{\delta x \, \delta y} = 2yz^2(2x) = 4xyz^2 \qquad \frac{\delta^2 f}{\delta z \, \delta y} = 2x^2y(2z) = 4x^2yz$$

$$\frac{\delta^2 f}{\delta x \, \delta z} = 2y^2z(2x) = 4xy^2z \qquad \frac{\delta^2 f}{\delta y \, \delta z} = 2x^2z(2y) = 4x^2yz$$

■■■ **3-27**

Let D_i denote partial differentiation with respect to the ith variable.
If $f(x,y,z) = x^2y^2z^3 + 3x \sin z$ then $D_{123}f$ is (a) $2 + 3 \cos z$
(b) $12xyz^2$ (c) $12xyz^2 + \cos z$ (d) $12xyz^2 - \cos z$ (e) $4xyz^3$.

****

$$D_1 f = 2xy^2z^3 + 3\sin z \, , \quad D_{12}f = 4xyz^3 , \; D_{123}f = 12xyz^2.$$

3-28

Find f_{xx}, f_{yy} and f_{yx} if $f(x,y) = \sin x^2 y$.

****

$$f(x,y) = \sin x^2 y$$

$$f_x(x,y) = 2xy \cos x^2 y$$

$$f_{xx}(x,y) = 2y \cos x^2 y - 4x^2 y^2 \sin x^2 y$$

$$f_y(x,y) = x^2 \cos x^2 y$$

$$f_{yy}(x,y) = -x^4 \sin x^2 y$$

$$f_{xy} = f_{yx} = 2x \cos x^2 y - x^4 \sin x^2 y.$$

3-29

If $f(x,y,z) = x \ln(yz^2)$ find $f_{xy}(x,y,z)$, $f_{xz}(x,y,z)$ and $f_{yz}(x,y,z)$

****

$$f_x(x,y,z) = \ln(yz^2)$$

$$f_{xy}(x,y,z) = f_y[\ln(yz^2)] = \frac{1}{yz^2} \cdot z^2 = \frac{1}{y}$$

$$f_{xz}(x,y,z) = f_z[\ln(yz^2)] = \frac{1}{yz^2} \cdot 2z = \frac{2}{yz}$$

$$f_y(x,y,z) = x \cdot \frac{1}{yz^2} \cdot z^2 = \frac{x}{y}$$

$$f_{yz}(x,y,z) = f_z\left[\frac{x}{y}\right] = 0$$

━ 3-30

If $z = x^2 \sin y + ye^x$,

Find $\dfrac{\partial z}{\partial x}, \dfrac{\partial z}{\partial y}, \dfrac{\partial^2 z}{\partial x^2}, \dfrac{\partial^2 z}{\partial x \partial y}$ and $\dfrac{\partial^2 z}{\partial x^2}.$

****

$$\frac{\partial z}{\partial x} = 2x \sin y + y e^x$$

$$\frac{\partial^2 z}{\partial x^2} = 2 \sin y + y e^x$$

$$\frac{\partial^2 z}{\partial y \partial x} = 2x \cos y + e^x$$

$$\frac{\partial z}{\partial y} = x^2 \cos y + e^x$$

$$\frac{\partial^2 z}{\partial y^2} = -x^2 \sin y$$

$$\frac{\partial^2 z}{\partial x \partial y} = 2x \cos y + e^x$$

Note for a check: $\dfrac{\partial^2 z}{\partial x \partial y} = \dfrac{\partial^2 z}{\partial y \partial x}.$

3-31 ■■■

Let $f(x,y) = x^3y - xy^2 + y^4 + x.$ Find $\left.\dfrac{\partial^2 f}{\partial y\,\partial x}\right|_{(2,3)}$

****

$$\frac{\partial f}{\partial x} = 3x^2y - y^2 + 1$$

$$\left.\frac{\partial^2 f}{\partial y\,\partial x}\right|_{(2,3)} = \left.(3x^2 - 2y)\right|_{(2,3)} = 12 - 6 = 6$$

THE DIRECTIONAL DERIVATIVE

3-32 ■■■

Find the directional derivative of $f(x,y) = x^2 - 3xy + 2y^2$ at $(-1,2)$ in the direction of $i + \sqrt{3}j.$

****

THE UNIT VECTOR IN THE DIRECTION OF

$i + \sqrt{3}j$ IS $\dfrac{i + \sqrt{3}\,j}{2} = u$

$$\frac{\delta f}{\delta x} = 2x - 3y \, ; \quad AT \ (-1,2) \quad \frac{\delta f}{\delta x} = -2 - 6 = -8$$

$$\frac{\delta f}{\delta y} = -3x + 4y \, ; \quad AT \ (-1,2) \quad \frac{\delta f}{\delta y} = 3 + 8 = 11$$

$$\nabla f(-1,2) = -8i + 11j$$

$$f'_u(-1,2) = (-8i + 11j) \cdot \left(\frac{1}{2}i + \frac{\sqrt{3}}{2}j\right) = -4 + \frac{11\sqrt{3}}{2}$$

$$= \frac{11\sqrt{3} - 8}{2}$$

━━ **3-33**

Consider $f(x,y,z) = x^2y + y^3z + xz^3$ at the point $P = (2,1,-1)$.

(a) Find the directional derivative at P in the direction of $<1,2,3>$.

(b) Find a vector in the direction in which f increases most rapidly at P.

(c) Find the rate of change of f in the direction indicated in part (b).

$$**$$

(a) $f_x = 2xy + z^3, \quad f_y = x^2 + 3y^2z, \quad f_z = y^3 + 3xz^2$

$\vec{\nabla}f(2,1,-1) = \langle 3,1,7 \rangle, \quad \vec{u} = \frac{1}{\sqrt{14}}\langle 1,2,3 \rangle$

$D_{\vec{u}}f = \vec{\nabla}f \cdot \vec{u} = \langle 3,1,7 \rangle \cdot \frac{1}{\sqrt{14}}\langle 1,2,3 \rangle = \frac{26}{\sqrt{14}}$

(b) $\vec{\nabla}f(2,1,-1) = \langle 3,1,7 \rangle$

(c) $\|\vec{\nabla}f\| = \|\langle 3,1,7 \rangle\| = \sqrt{59}$

━━ **3-34**

For $f(x,y) = x^2y^3$, $\vec{u} = (3/5, -4/5)$, the directional derivative of f in the direction $\vec{u}$ at the point (x,y) is (a) $(6xy^3 - 12x^2y^2)/5$ (b) $(3x^2 - 4y^3)/5$ (c) $(6x - 12y^2)/5$ (d) $2xy^3 + 3x^2y^2$ (e) $\sqrt{4x^2y^6 + 9x^4y^4}$.

$$**$$

$\vec{\nabla}f = \langle 2xy^3, 3x^2y^2 \rangle, \quad D_{\vec{u}}f(x,y) = \vec{\nabla}f \cdot \vec{u} =$

$\langle 2xy^3, 3x^2y^2 \rangle \cdot \langle \frac{3}{5}, -\frac{4}{5} \rangle = 2xy^3 \cdot \frac{3}{5} + 3x^2y^2(-\frac{4}{5}) = \frac{6xy^3 - 12x^2y^2}{5}$

3-35 ■■

Find the directional derivative of $f = x^2 + y^2 - z$ at the point $(1,3,5)$ in the direction of $\vec{a} = 2\hat{i} - \hat{j} + 4\hat{k}$.

****

The directional derivative $= \dfrac{df}{ds} = \vec{\nabla} f \cdot \hat{u}$

$$\vec{\nabla} f = \left(\hat{i} \frac{\partial}{\partial x} + \hat{j} \frac{\partial}{\partial y} + \hat{k} \frac{\partial}{\partial z} \right)(x^2 + y^2 - z)$$

$$= \hat{i} 2x + \hat{j} 2y - \hat{k}$$

$$\hat{u} = \frac{\vec{a}}{a} = \frac{2\hat{i} - \hat{j} + 4\hat{k}}{\sqrt{(2)^2 + (-1)^2 + (4)^2}} = \frac{2\hat{i} - \hat{j} + 4\hat{k}}{\sqrt{21}}$$

$$\vec{\nabla} f \cdot \hat{u} = \left(\hat{i} 2x + \hat{j} 2y - \hat{k} \right) \cdot \left(\frac{2\hat{i} - \hat{j} + 4\hat{k}}{\sqrt{21}} \right)$$

$$= \frac{4x - 2y - 4}{\sqrt{21}}$$

At the point $(1, 3, 5)$, $\vec{\nabla} f \cdot \hat{u} = \dfrac{4(1) - 2(3) - 4}{\sqrt{21}}$

$$= -\frac{6}{\sqrt{21}} = -\frac{2\sqrt{21}}{7}$$

Therefore, $\dfrac{df}{ds} = -\dfrac{2\sqrt{21}}{7}$ at $(1, 3, 5)$.

■■■ **3-36**

Let the temperature in a flat plate be given by the function

$$T(x,y) = 3x^2 + 2xy$$

What is the value of the directional derivative of this function at the point (3,−6) in the direction $v = 4\vec{i} - 3\vec{j}$? In what direction is the plate cooling most rapidly?

****

Let $\vec{u}$ = unit directional vector

$$= \frac{\vec{v}}{|\vec{v}|} = \frac{4}{5}\vec{i} - \frac{3}{5}\vec{j}$$

$$\nabla T = (6x+2y)\vec{i} + 2x\vec{j}$$

$$\nabla T \cdot \vec{u} = \frac{4}{5}(6x+2y) - \frac{3}{5}(2x)$$

$$= \frac{18}{5}x + \frac{8}{5}y \Big|_{(3,-6)} = \frac{6}{5}$$

$\therefore$ $\frac{6}{5}$ is the value of the directional derivative of $T(x,y)$ at $(3,-6)$ in the direction $4\vec{i} - 3\vec{j}$.

Also, $\nabla T(3,-6) = 6\vec{i} + 6\vec{j}$

$\therefore$ The plate is <u>cooling</u> most rapidly in the direction $-\vec{i} - \vec{j}$ from the point $(3,-6)$.

3-37 ■■

Find the directional derivative of $f(x,y) = x^3 + xy^2$ at the point $(1,-2)$ in the direction toward the origin.

****

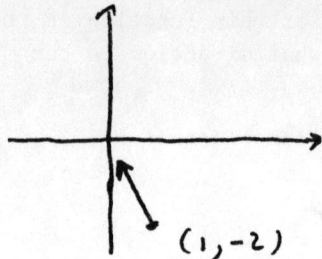

$$\frac{df}{ds} = (grad f) \cdot \frac{\underline{V}}{\|\underline{V}\|}$$

$$\underline{V} = (0,0) - (1,-2) = (-1,2)$$

$$grad f = (3x^2 + y^2, 2xy)$$

at $(1,-2)$

$$grad f = (7, -4)$$

so $$\frac{df}{ds} = \frac{(7,-4) \cdot (-1,2)}{\sqrt{1+4}} = \frac{-15}{\sqrt{5}} = -3\sqrt{5}$$

3-38 ■■■■■■■■■■■■■■■■■■■■■■■■■■■■■■■■■■■■■■■

Find the directional derivative of $f(x,y) = 3x^2 + xy - y^3$ in the direction $\pi/3$.

****

The directional derivative is $f_x(x,y) \cos(\theta) + f_y(x,y) \sin(\theta)$ and U is the unit vector in the direction θ.

$$\therefore D_{\underline{u}} f(x,y) = (6x+y) \cos\left(\frac{\pi}{3}\right) + (x - 3y^2) \sin\left(\frac{\pi}{3}\right)$$

$$= (6x+y)\left(\frac{1}{2}\right) + (x-3y^2)\left(\frac{\sqrt{3}}{2}\right)$$

$$= 3x + \frac{1}{2}y + \frac{\sqrt{3}}{2}x - \frac{3\sqrt{3}}{2}y^2$$

$$= \left(\frac{6+\sqrt{3}}{2}\right)x + \frac{1}{2}y - \frac{3\sqrt{3}}{2}y^2$$

3-39

Let $f(x, y, z) = x^2 + y^2 + xz$. Find the directional derivative of f at (1, 2, 0) in the direction of the vector v= (1, -1, 1).

****

The gradient is given by $\vec{\nabla f}(x,y,z) = (2x+z, 2y, x)$

Hence $\vec{\nabla f}(1,2,0) = (2,4,1)$

The directional derivative at (x,y,z) in direction of V is given by

$$D_v(x,y,z) = \vec{\nabla f}(x,y,z) \cdot \frac{V}{\text{length of } V} ; \text{ hence,}$$

$$D_v(1,2,0) = (2,4,1) \cdot \left(\frac{1}{\sqrt{3}}, -\frac{1}{\sqrt{3}}, \frac{1}{\sqrt{3}}\right)$$

$$= \frac{2}{\sqrt{3}} - \frac{4}{\sqrt{3}} + \frac{1}{\sqrt{3}}$$

$$= -\frac{1}{\sqrt{3}}$$

3-40

Find the directional derivative of $f(x,y,z) = x^3 + y^2 - z$ at the point (1,-1,2) in the direction of the vector (-1,2,2).

****

$$\vec{N} = \frac{(-1,2,2)}{\|(-1,2,2)\|} = \left(-\frac{1}{3}, \frac{2}{3}, \frac{2}{3}\right)$$

$$\nabla f \cdot \vec{N} = (3x^2, 2y, -1)\Big|_{(1,-1,2)} \cdot \left(-\frac{1}{3}, \frac{2}{3}, \frac{2}{3}\right)$$

$$= (3, -2, -1) \cdot \left(-\frac{1}{3}, \frac{2}{3}, \frac{2}{3}\right) = -1 - \frac{4}{3} - \frac{2}{3}$$

$$= -3$$

3-41 ■■■

Calculate the directional derivative of the function $f(x,y) = 4 - x^3 - y^3$ at the point $(1,1)$ in the direction equidistant between x and y axes in the first quadrant.

****

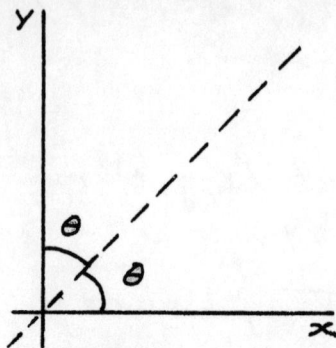

The direction "equidistant between x and y axes in the first quadrant" in the x-y plane is shown alongside, from which we note that the unit vector is given by :

$$\vec{u} = \frac{1}{\sqrt{2}} (1,1)$$

For calculating the directional derivative we let $P_0 = (1,1)$ and $P = (x,y)$, so that the vector equation $\overrightarrow{P_0 P}$ is given by :

$$\overrightarrow{P_0 P} = t \cdot u = (x-1, y-1)$$
$$= \frac{t}{\sqrt{2}} (1,1)$$

Hence, $x = 1 + \frac{t}{\sqrt{2}}$, and $y = 1 + \frac{t}{\sqrt{2}}$

We now determine $f(P) - f(P_0)$ by writing

$$f(P) - f(P_0) = f(x,y) - f(1,1)$$
$$= (4 - x^3 - y^3) - f(1,1)$$
$$= 4 - \left(1 + \frac{t}{\sqrt{2}}\right)^3 - \left(1 + \frac{t}{\sqrt{2}}\right)^3 - 2$$
$$= 4 - \left[1 + 3\frac{t}{\sqrt{2}} + 3\left(\frac{t}{\sqrt{2}}\right)^2 + \left(\frac{t}{\sqrt{2}}\right)^3\right]$$
$$- \left[1 + 3\frac{t}{\sqrt{2}} + 3\left(\frac{t}{\sqrt{2}}\right)^2 + \left(\frac{t}{\sqrt{2}}\right)^3\right] - 2$$

$$= 4 - 1 - 3\frac{t}{\sqrt{2}} - \frac{3t^2}{2} - \frac{t^3}{2\sqrt{2}} - 1 - 3\frac{t}{\sqrt{2}} - \frac{3t^2}{2} - \frac{t^3}{2\sqrt{2}} - 2$$

$$= -2\left(\frac{3t}{\sqrt{2}}\right) - 2\left(\frac{3t^2}{2}\right) - 2\left(\frac{t^3}{2\sqrt{2}}\right)$$

$$= -3\sqrt{2}\, t - 3t^2 - \frac{t^3}{\sqrt{2}}$$

Hence, we write :

$$D_u \ f \ (1,1) = \lim_{t \to 0} \frac{f(P) - f(P_0)}{t}$$

$$= \lim_{t \to 0} \frac{-3\sqrt{2} \ t - 3t^2 - \frac{t^3}{\sqrt{2}}}{t}$$

$$= \lim_{t \to 0} -3\sqrt{2} - 3t - \frac{t^2}{\sqrt{2}}$$

$$= -3\sqrt{2} \qquad \underline{Ans.}$$

━━━━━━━━━━━━━━━━━━━━━━━━━━━━━━━━**3-42**

Find the directional derivative of the function $f(x,y) = x^2 y^5 + x^3$ at the point $(-1,2)$ in the direction toward $(2,3)$.

****

A VECTOR IN THE DIRECTION FROM $(-1,2)$ TO $(2,3)$
IS
$$[2-(-1), \ 3-2] = [3, 1].$$
THE UNIT VECTOR IN THAT DIRECTION IS
$$\frac{[3, 1]}{\sqrt{3^2 + 1^2}} = \left[\frac{3}{\sqrt{10}}, \frac{1}{\sqrt{10}}\right].$$

THE PARTIAL DERIVATIVES ARE
$$\frac{\partial f}{\partial x} = 2xy^5 + 3x^2$$
$$\frac{\partial f}{\partial y} = 5x^2 y^4,$$
AT $(-1,2)$
$$\frac{\partial f}{\partial x}(-1,2) = 2(-1)2^5 + 3(-1)^2 = -64 + 3 = -61$$
$$\frac{\partial f}{\partial y}(-1,2) = 5(-1)^2 2^4 = 80.$$

THE DIRECTIONAL DERIVATIVE IS
$$-61\left(\frac{3}{\sqrt{10}}\right) + 80\left(\frac{1}{\sqrt{10}}\right) = -\frac{103}{\sqrt{10}}.$$

3-43 ■■■

The Intergalactic Explorer ship Zora is hanging motionless at (5,0,10) (Universal Galactic Coordinates) when the crew spots an interesting object at (7,5,6). The temperature in that part of the galaxy is given by

$$T(x,y,z) = x^2 + y/x + z^3 .$$

As the crew starts to move the Zora directly toward the unknown object, what is the rate of change of the temperature (in degrees/Universal Galactic Units)?

****

The solution to this problem is the directional derivative $D_{\vec{u}} T (5,0,10)$, where $\vec{u}$ is the unit vector in the direction of the Zora's motion.

A vector in the direction of the ship's motion is

$(7,5,6) - (5,0,10) = (2,5,-4)$, so

$$\vec{u} = \frac{(2,5,-4)}{\|(2,5,-4)\|} = \frac{1}{\sqrt{45}} (2,5,-4)$$

$$\nabla T \Big|_{(5,0,10)} = (2x - \frac{y}{x^2}, \frac{1}{x}, 3z^2) \Big|_{(5,0,10)} = (10, \frac{1}{5}, 300)$$

$$D_{\vec{u}} T (5,0,10) = \nabla T \Big|_{(5,0,10)} \cdot \vec{u} = (10, \frac{1}{5}, 300) \cdot \frac{(2,5,-4)}{\sqrt{45}}$$

$$= \frac{-1179}{\sqrt{45}}$$

The temperature is _decreasing_ at $\frac{1179}{\sqrt{45}}°$/ Univ. Gal. Unit the instant the Zora leaves the point (5,0,10).

■■**3-44**

Let $f(x,y) = x^2 y + \ln y \quad (y>0)$

Find the directional derivative of $f(x,y)$ at the point $(1,1)$ in the direction of the origin.

****

the directional derivative $D_{\vec{u}} f(x,y) = f_x(x,y) u_1 + f_y(x,y) u_2$

where $u_1 \vec{\imath} + u_2 \vec{\jmath}$ is a <u>unit</u> vector in the desired direction.

$f_x = 2xy \qquad$ so $\qquad f_x(1,1) = 2$

$f_y = x^2 + \frac{1}{y} \qquad$ so $\qquad f_y(1,1) = 2.$

Let $\vec{v} = $ vector from $(1,1)$ to $(0,0)$

then $\vec{v} = -\vec{\imath} - \vec{\jmath}$

Since $\|v\| = \sqrt{2}$, the unit vector in this direction is: $\vec{u} = -\frac{1}{\sqrt{2}} \vec{\imath} - \frac{1}{\sqrt{2}} \vec{\jmath}$

Hence $D_{\vec{u}} f(1,1) = f_x u_1 + f_x u_2 = 2\left(-\frac{1}{\sqrt{2}}\right) + 2\left(-\frac{1}{\sqrt{2}}\right)$

$$= -\frac{2}{\sqrt{2}} = -\sqrt{2}.$$

3-45 ■■■

A bug is crawling on the surface $z = x^2 + xy + 2y^2$. When he reaches the point (2, 1, 8) he wants to avoid vertical change. In which direction should he head? [He wants the directional derivative to be zero.]

****

$\nabla z = \langle 2x+y, x+4y \rangle$, so $\nabla z(2,1) = \langle 5,6 \rangle$

Since $\dfrac{\partial z}{\partial u} = \nabla z \cdot \vec{u}$ for any unit vector

$\vec{u}$ and he wants $\dfrac{\partial z}{\partial u} = 0$, choose

$\vec{u} = \pm \dfrac{\langle -6, 5 \rangle}{\sqrt{61}}$.

DIFFERENTIABILITY
AND THE TOTAL DIFFERENTIAL

3-46 ■■■

A right triangle has leg A with length 4, leg B with length 3, and hypotenuse with length 5. Use a total differential to approximate the length of the hypotenuse if leg A had length 4.2 and leg B had length 2.9 .

****

The function relating A, B and h is $h = \sqrt{A^2 + B^2}$.

$dh = \frac{1}{2}(A^2+B^2)^{-1/2} 2A\, dA + \frac{1}{2}(A^2+B^2)^{-1/2} 2B\, dB$

$\qquad = (A^2+B^2)^{-1/2} A\, dA + (A^2+B^2)^{-1/2} B\, dB$

$A = 4, \quad B = 3, \quad dA = .2, \quad dB = -.1$

$dh = (4^2+3^2)^{-1/2}(4)(.2) + (4^2+3^2)^{-1/2}(3)(-.1)$

$\qquad = 25^{-1/2}(.8) + 25^{-1/2}(-.3) = .16 - .06 = .10$

$\therefore$ hypotenuse is approximately $5 + .10 = 5.10$

━━━━━━━━━━━━━━━━━━━━━━━━━━━━━━━━━━━━━━ **3-47**

Suppose $f(1,1) = 2$ and $\nabla f(0,0) = 2\vec{i} - \vec{j}$
Use differentials to estimate $f(0.8, 1.1)$

****

Recall: $\nabla f = f_x \vec{i} + f_y \vec{j}$

Hence $f_x (1,1) = 2$ and $f_y (1,1) = -1$

$\Delta f \approx df = f_x\, dx + f_y\, dy$

$\qquad = 2(-0.2) + (-1)(0.1) = -0.5$

Hence: $f(0.8, 1.1) \approx f(0,0) + df = 2 - 0.5$

$\qquad\qquad\qquad\qquad\qquad\qquad = 1.5$

━━━━━━━━━━━━━━━━━━━━━━━━━━━━━━━━━━━━━━ **3-48**

Let $f(x,y) = x^2 y$, $P_0 = (x_0, y_0)$, $P = (x_0 + \Delta x, y_0 + \Delta y)$. Then

$\Delta f - df$ is equal to (a) $(2x_0 \Delta y + y_0 \Delta x + \Delta x \Delta y) \Delta x$ (b) $2y_0 \Delta x + x_0 \Delta y$
(c) $2x_0 \Delta y + y_0 \Delta x + \Delta x \Delta y) \Delta x + x_0^2 y_0$ (d) $x_0^2 \Delta y + (2x_0 y_0 + 2x_0 \Delta y) \Delta x$
(e) $(\Delta y + x_0 y_0) \Delta x + 2x_0 \Delta y$.

****

$\Delta f - df = \overbrace{f(x_0 + \Delta x, y_0 + \Delta y) - f(x_0, y_0)}^{\Delta f} - \overbrace{\left[f_x(x_0, y_0)\Delta x + f_y(x_0, y_0)\Delta y\right]}^{df}$

$= (x_0 + \Delta x)^2 (y_0 + \Delta y) - x_0^2 y_0 - 2x_0 y_0 \Delta x - x_0^2 \Delta y$

$= (x_0^2 + 2x_0 \Delta x + \Delta x^2)(y_0 + \Delta y) - x_0^2 y_0 - 2x_0 y_0 \Delta x - x_0^2 \Delta y$

$= \cancel{x_0^2 y_0} + \cancel{x_0^2 \Delta y} + \cancel{2x_0 y_0 \Delta x} + 2x_0 \Delta x \Delta y + y_0 \Delta x^2 + \Delta x^2 \Delta y$
$\qquad\qquad\qquad - \cancel{x_0^2 y_0} - \cancel{2x_0 y_0 \Delta x} - \cancel{x_0^2 \Delta y}$

$= 2x_0 \Delta x \Delta y + y_0 \Delta x^2 + \Delta x^2 \Delta y = (2x_0 \Delta y + y_0 \Delta x + \Delta x \Delta y) \Delta x.$

3-49 ■■

Use the definition of differentiability to show that $f(x,y) = x^2 + y^2$ is differentiable at (6,8).

****

$$f(\vec{x_0}) = f(6,8) = 36 + 64 = 100$$

$$\frac{\partial f}{\partial x}\bigg|_{(6,8)} = 2x\bigg|_{x=6} = 12 \quad ; \quad \frac{\partial f}{\partial y}\bigg|_{(6,8)} = 2y\bigg|_{y=8} = 16$$

$$\ell(\vec{x}-\vec{x_0}) = \left(\frac{\partial f(\vec{x_0})}{\partial x}, \frac{\partial f(\vec{x_0})}{\partial y}\right)\cdot(\vec{x}-\vec{x_0}) = (12,16)\cdot(x-6,y-8)$$

$$= 12(x-6) + 16(y-8)$$

$$f(\vec{x}) = f(\vec{x_0}) + \ell(\vec{x}-\vec{x_0}) + e(\vec{x})$$

$$x^2+y^2 = 100 + 12(x-6) + 16(y-8) + e(\vec{x})$$

$$\therefore \quad e(\vec{x}) = x^2+y^2 - 12x - 16y + 100$$

$$\lim_{\vec{x}\to\vec{x_0}} \frac{e(\vec{x})}{|\vec{x}-\vec{x_0}|} = \lim_{(x,y)\to(6,8)} \frac{x^2+y^2-12x-16y+100}{\sqrt{(x-6)^2+(y-8)^2}}$$

$$= \lim_{(x,y)\to(6,8)} \frac{(x-6)^2+(y-8)^2}{\sqrt{(x-6)^2+(y-8)^2}}$$

$$= \lim_{(x,y)\to(6,8)} \sqrt{(x-6)^2+(y-8)^2} = 0$$

$$\therefore \quad f(x,y) = x^2+y^2 \quad \text{IS DIFFERENTIABLE AT } (6,8)$$

$$\text{BECAUSE} \quad \lim_{\vec{x}\to\vec{x_0}} \frac{e(\vec{x})}{|\vec{x}-\vec{x_0}|} = 0.$$

■■ 3-50

Find the total differential of w if $w = f(x,y,z) = xy^2z^3 - e^{xz}y$

****

The total differential of w is expressed as

$$dw = \frac{\partial w}{\partial x}\, dx + \frac{\partial w}{\partial y}\, dy + \frac{\partial w}{\partial z}\, dz,$$

where

$$\frac{\partial w}{\partial x} = y^2z^3 - yze^{xz}\ \ and$$

$$\frac{\partial w}{\partial y} = 2xyz^3 - e^{xz}\ \ and$$

$$\frac{\partial w}{\partial z} = 3xy^2z^2 - xye^{xz}.$$

$$\therefore\ \ dw = (y^2z^3 - yze^{xz})dx + (2xyz^3 - e^{xz})dy$$
$$+ (3xy^2z^2 - xye^{xz})dz$$

■■ 3-51

Suppose $f(x,y,z) = x^2y + xz^2 - y^3z$. Use the total differential to estimate the change in f as (x,y,z) varies from $(2,-1,3)$ to $(2.01,-0.97,2.98)$.

****

Let $w = f(x,y,z)$.

$f_x = 2xy + z^2$, $\quad f_y = x^2 - 3y^2z$, $\quad f_z = 2xz - y^3$

$f_x(2,-1,3) = 5$, $\quad f_y(2,-1,3) = -5$, $\quad f_z(2,-1,3) = 13$

$\Delta w \approx dw = \ \ 5\, dx - 5\, dy + 13\, dz$

$$= \ \ 5(0.01) - 5(0.03) + 13(-0.02) = -0.36$$

3-52 ■■

Use the total differential to approximate $\sqrt[3]{25}\ \sqrt[4]{17}$.

****

LET $f(x,y) = \sqrt[3]{x}\ \sqrt[4]{y} = x^{1/3}\ y^{1/4}$.
THEN THE TOTAL DIFFERENTIAL IS

$$df = \frac{1}{3} x^{-2/3} y^{1/4} dx + \frac{1}{4} x^{1/3} y^{-3/4} dy.$$

WE ASSUME THAT

$$f(x+dx, y+dy) \doteq f(x,y) + \frac{1}{3} x^{-2/3} y^{1/4} dx + \frac{1}{4} x^{1/3} y^{-3/4} dy.$$

AT $x=27$, $y=16$, $dx=-2$, AND $dy=1$,

$$f(27-2, 16+1) \doteq f(27,16) + \frac{1}{3} \cdot 27^{-2/3} 16^{1/4}(-2) + \frac{1}{4} \cdot 27^{1/3} 16^{-3/4} \cdot 1$$

$$= 3 \cdot 2 + \frac{1}{3} \cdot \frac{1}{9} \cdot 2 \cdot (-2) + \frac{1}{4} \cdot 3 \cdot \frac{1}{8}$$

$$= 6 - \frac{4}{27} + \frac{3}{32}$$

$$= 6 - \frac{47}{864}$$

$$= 5\frac{817}{864}.$$

THEREFORE, $\sqrt[3]{25}\ \sqrt[4]{17} \doteq 5\frac{817}{864} \doteq 5.95$.

IMPLICIT FUNCTIONS

■■■ **3-53**

Find dy/dx if $x\cos(y) = \ln(z)$ and $\sin(x) = ze^y$.

****

$$F(x, y, z) = x\cos y - \ln z \qquad G(x, y, z) = ze^y - \sin x$$

$$\frac{dy}{dx} = -\frac{J\left(\frac{F, G}{X, Z}\right)}{J\left(\frac{F, G}{Y, Z}\right)} = -\frac{\begin{vmatrix} F_x & F_z \\ G_x & G_z \end{vmatrix}}{\begin{vmatrix} F_y & F_z \\ G_y & G_z \end{vmatrix}} = -\frac{F_x G_z - F_z G_x}{F_y G_z - F_z G_y}$$

$$= -\frac{(\cos y)(e^y) - \left(-\frac{1}{z}\right)(-\cos x)}{(-x\sin y)e^y - \left(-\frac{1}{z}\right)(ze^y)}$$

$$= \frac{ze^y \cos y - \cos x}{xze^y \sin y - ze^y}$$

3-54 ■■■

Determine by means of the implicit function theorem if the equation $x^3 + y - y^3 = 0$ can be used to obtain y as a function of x locally about each point (x_0, y_0) on the graph.

****

We note that the given equation can be written in the form : $f(x,y) = 0$.

By using the notation $D_2 f$ as the partial derivative of f with respect to its second variable, we need to find if $D_2 f(x,y)$ is ever equal to zero along the level set.

From the given equation we write for the partial derivative :

$$D_2 f(x,y) = 1 - 3y^2$$

Now, in order to make $D_2 f(x,y) = 0$ we have to set $y^2 = \frac{1}{3}$

OR

$$y = \pm \frac{1}{\sqrt{3}}$$

By substituting the value of y in the given level set equation we get

$$x^3 \pm \left[\left(\frac{1}{\sqrt{3}}\right) - \left(\frac{1}{\sqrt{3}}\right)^3 \right] = 0$$

OR

$$x^3 \pm \left(\frac{1}{\sqrt{3}} - \frac{1}{3\sqrt{3}} \right) = 0$$

$$x = \mp \left[\frac{3-1}{3\sqrt{3}} \right]^{1/3} = \mp \frac{2^{1/3}}{\sqrt{3}}$$

Hence, we find that (i) there are two points, $\pm \left[\frac{2^{1/3}}{\sqrt{3}}, -\frac{1}{\sqrt{3}} \right]$, one each for x and y , at which $D_2 f = 0$, and (ii) at every other point (x_0, y_0) on the level set the second partial derivative $D_2 f$ is not zero.

Therefore, we conclude that the level set is a function graph near the point (x_0, y_0).

■■■ **3-55**

Use implicit differentiation to find $\frac{\partial z}{\partial x}$ on the surface given by $x^3y + y^2z^2 + xz^3 = 3$.

****

Differentiate w.r.t. x (holding y constant):

$$3x^2y + 2y^2z\frac{\partial z}{\partial x} + \left(z^3 + 3xz^2\frac{\partial z}{\partial x}\right) = 0$$

$$2y^2z\frac{\partial z}{\partial x} + 3xz^2\frac{\partial z}{\partial x} = -(3x^2y + z^3)$$

$$\frac{\partial z}{\partial x} = -\frac{3x^2y + z^3}{2y^2z + 3xz^2}$$

■■■ **3-56**

If $z^3 + xz - y = 0$, find $\frac{\partial^2 z}{\partial x \partial y}$ in terms of x, y and z.

****

$$3z^2\frac{\partial z}{\partial y} + x\frac{\partial z}{\partial y} - 1 = 0 \qquad 3z^2\frac{\partial z}{\partial x} + x\frac{\partial z}{\partial x} + z = 0$$

$$\therefore \frac{\partial z}{\partial y} = \frac{1}{3z^2 + x} \qquad \qquad \therefore \frac{\partial z}{\partial x} = \frac{-z}{3z^2 + x}$$

$$\frac{\partial^2 z}{\partial x \partial y} = \frac{-1\left(6z\frac{\partial z}{\partial x} + 1\right)}{(3z^2 + x)^2} = \frac{-1\left(\frac{-6z^2}{3z^2 + x} + 1\right)}{(3z^2 + x)^2} = \frac{3z^2 - x}{(3z^2 + x)^3}$$

CHAIN RULES AND THE GRADIENT

3-57 ■■

One side of a rectangle is increasing at 4 ft/min and the other at
7 ft/min. At the time when the first side is 24 ft and the second is
32 ft, find:

a. how fast the area is changing.
b. how fast the diagonal is changing.

****

We call the length of the first side x and the length of the second side y. Thus we are given that $x = 24$, $y = 32$, $\frac{dx}{dt} = 4$, and $\frac{dy}{dt} = 7$.

a. Let the area be A. Then $A = xy$

Using the Chain Rule: $\frac{dA}{dt} = y \frac{dx}{dt} + x \frac{dy}{dt}$

$$= 32(4) + 24(7)$$
$$= 296 \text{ square ft/min}$$

b. Let the diagonal be P. Then $P = \sqrt{x^2 + y^2}$.

As in part a, $\frac{dP}{dt} = \frac{x}{\sqrt{x^2+y^2}} \frac{dx}{dt} + \frac{y}{\sqrt{x^2+y^2}} \frac{dy}{dt}$

$$= \frac{24}{\sqrt{24^2+32^2}} (4) + \frac{32}{\sqrt{24^2+32^2}} (7)$$

$$= \frac{24}{40} (4) + \frac{32}{40} (7) = \frac{3}{5} (4) + \frac{4}{5} (7)$$

$$= 8 \text{ ft/min}$$

3-58

You are standing at the point (1,2,12) on a hillside which has equation
$f(x,y) = x^2y - xy + y^3 + xy^2$. Find a two-dimensional vector which points
the direction you should face to go up the hill as fast as possible, and
find the slope in that direction.

****

The gradient vector points the direction for greatest
increase. $\vec{\nabla}f(x,y) = (2xy - y + y^2)\vec{i} + (x^2-x+3y^2+2xy)\vec{j}$

$\vec{\nabla}f(1,2) = (2\cdot1\cdot2 - 2 + 2^2)\vec{i} + (1^2-1+3\cdot2^2+2\cdot1\cdot2)\vec{j}$

$= (4-2+4)\vec{i} + (1-1+12+4)\vec{j} = 6\vec{i} + 16\vec{j}$

This is the direction to face to climb the hill as fast
as possible.

The slope in the direction of the greatest increase is the
magnitude of the gradient. $|\vec{\nabla}f(1,2)| = \sqrt{6^2 + 16^2}$

$= \sqrt{36+256} = \sqrt{292} = 2\sqrt{73}$.

3-59

Given $f(x,y) = x^2y$, $P_o = (3,2)$, and let $\vec{u}$ be the unit vector for which
the directional derivative $D_{\vec{u}} f(P_o)$ has maximum value. This maximum
value is (a) 288/5 (b) 1 (c) 6 (d) $12\sqrt{3}$ (e) 15.

****

This value is $|\vec{\nabla}f(P_o)|$. $\vec{\nabla}f = \langle 2xy, x^2\rangle$, $\vec{\nabla}f(P_o)=\langle12,9\rangle$

$|\vec{\nabla}f(P_o)| = \sqrt{144+81} = \sqrt{225} = 15$.

3-60

Let $f(x,y) = 3x^3 + y^2 - 9x + 4y$. Find the direction at the origin in which $f(x,y)$ is <u>decreasing</u> the fastest.

****

Recall: the direction of fastest <u>increase</u> of f at (x,y) is given by $\nabla f(x,y)$

$\therefore$ the direction of fastest <u>decrease</u> must be the opposite direction, i.e. $-\nabla f(x,y)$

$f_x = 9x^2 - 9 \qquad$ so $f_x(0,0) = -9$

$f_y = 2y + 4 \qquad$ so $f_y(0,0) = 4$

Hence $-\nabla f(0,0) = -(-9\vec{i} + 4\vec{j}) = 9\vec{i} - 4\vec{j}$

3-61

Suppose that $z = u^2 + uv + v^3$, and that $u = 2x^2 + 3xy$ and $v = 2x - 3y + 2$. Find $\frac{\partial z}{\partial x}$ at $(x,y) = (1,2)$.

****

Note: when $(x,y) = (1,2)$, $u = 8$ and $v = -2$

$\frac{\partial z}{\partial x} = \frac{\partial z}{\partial u} \frac{\partial u}{\partial x} + \frac{\partial z}{\partial v} \frac{\partial v}{\partial x}$

$= (2u + v)(4x + 3y) + (u + 3v^2)(2)$

Evaluating at $(x,y) = (1,2)$ and $(u,v) = (8,-2)$

gives $\quad (14)(10) + (20)(2) = 180$.

■■■ **3-62**

Determine if the vector $(x - 2y)i - 2xj$ is a gradient of the function f. If it is, find $f(x,y)$

****

If the vector is a gradient, $f_{xy}(x,y) = f_{yx}(x,y)$ where $f_x(x,y) = x - 2y$ and $f_y(x,y) = -2x$

$f_{xy}(x,y) = f_y(x-2y) = -2$ and $f_{yx}(x,y) = f_x(-2x) = -2$

So the vector is $\nabla f(x,y)$. Since $f_x(x,y) = x-2y$, integrating both sides with respect to x gives

$$f(x,y) = \tfrac{1}{2}x^2 - 2xy + T(y) \text{ where } T(y) \text{ is}$$

independent of x. Taking the partial derivative of this equation with respect to y gives

$$f_y(x,y) = -2x + T'(y). \text{ But } f_y(x,y) = -2x$$

$$\text{So } T'(y) = 0 \text{ and } T(y) = C$$

Substituting back for $T(y)$, the equation for $f(x,y)$ is

$$f(x,y) = \tfrac{1}{2}x^2 - 2xy + C$$

■■■ **3-63**

Given $f = x^2y^3$, $x = u^2 + v^2$, $y = 2u + 3v$, the partial derivative of f with respect to v is (a) $(2x + 3y^2)(2v + 3)$ (b) $xy^2(4yv + 9x)$ (c) $2(xy^3 + v) + 3(x^2y^2 + 1)$ (d) $2vy^3 + 3x^2$ (e) $4xy^3u + 6x^2y^2$.

****

By the chain rule, $\dfrac{\partial f}{\partial v} = \dfrac{\partial f}{\partial x} \cdot \dfrac{\partial x}{\partial v} + \dfrac{\partial f}{\partial y} \cdot \dfrac{\partial y}{\partial v} = $

$2xy^3(2v) + 3x^2y^2(3) = 4xy^3v + 9x^2y^2 = xy^2(4yv + 9x).$

3-64 ■■■

If f is a function of x and y, and y is a function of x then indirectly f depends only on x: "f"(x) = f(x, y(x)).

a) Use the chain rule to write an expression for d"f"/dx in terms of ∂f/∂x and ∂f/∂y.

b) If $f(x,y) = \sin x + \sqrt{1 - y^2}$ and y(x) = cos x calculate d"f"/dx in two different ways:
   1. substitute for y and calculate "f"(x) directly, then differentiate.
   2. use the formula you got in part a).

****

a) $\dfrac{d\,"f"}{dx} = \dfrac{\partial f}{\partial x}\dfrac{dx}{dx} + \dfrac{\partial f}{\partial y}\dfrac{dy}{dx} = \boxed{\dfrac{\partial f}{\partial x} + \dfrac{\partial f}{\partial y}\dfrac{dy}{dx}}$

b-1) $"f"(x) = f(x, \cos x) = \sin x + \sqrt{1 - \cos^2 x} = 2\sin x$

So $\boxed{\dfrac{d\,"f"}{dx} = 2\cos x}$

b-2) Since $\dfrac{\partial f}{\partial x} = \cos x$ and $\dfrac{\partial f}{\partial y} = \dfrac{-y}{\sqrt{1-y^2}}$

and $\dfrac{dy}{dx} = -\sin x$, we substitute these expressions

into the formula for part (a) to get

$\dfrac{d\,"f"}{dx} = \cos x + \left(\dfrac{-y}{\sqrt{1-y^2}}\right)(-\sin x)$

But our answer should only involve x, so we substitute y(x) = cos x to get

$\dfrac{d\,"f"}{dx} = \cos x + \left(\dfrac{-\cos x}{\sqrt{1-\cos^2 x}}\right)(-\sin x)$

$= \cos x + \dfrac{\cos x}{\sin x}\cdot \sin x$

$\boxed{= 2\cos x}$, which agrees — of course — with our previous answer

Note: in light of the work in b-2, a more correct answer to (a) would have been: $\dfrac{d\,"f"}{dx} = \dfrac{\partial f}{\partial x}(x, y(x)) + \dfrac{\partial f}{\partial y}(x, y(x))\dfrac{dy}{dx}$

■■■ **3-65**

Given $f(x,y) = (1/4)x^4 y^3$, at the point $(-1,2)$,

    a. find the directional derivative's maximum value,

    b. find the unit vector in the direction in which the directional derivative takes on its maximum value.

****

a. THE DIRECTIONAL DERIVATIVE'S MAXIMUM VALUE IS THE LENGTH (MAGNITUDE) OF THE GRADIENT VECTOR, $\vec{\nabla} f$. THE GRADIENT IS

$$\vec{\nabla} f(x,y) = \left[x^3 y^3, \ \frac{3}{4} x^4 y^2 \right]$$

$$\vec{\nabla} f(-1,2) = \left[(-1)^3 \cdot 2^3, \ \frac{3}{4}(-1)^4 \cdot 2^2 \right] = [-8, 3].$$

SO THE DIRECTIONAL DERIVATIVE'S MAXIMUM VALUE IS

$$\sqrt{(-8)^2 + 3^2} = \sqrt{73}.$$

b. THE DIRECTION OF MAXIMUM VALUE FOR THE DIRECTIONAL DERIVATIVE IS THE DIRECTION OF THE GRADIENT, $[-8, 3]$. THE UNIT VECTOR IN THIS DIRECTION IS

$$\frac{[-8, 3]}{\sqrt{73}} = \left[\frac{-8}{\sqrt{73}}, \ \frac{3}{\sqrt{73}} \right].$$

3-66

If $f(x,y) = 2x^2 + 4y^2 - xy$, find the gradient at the point (2,1). Also find the rate of change of $f(x,y)$ in the direction $\pi/3$ at (2,1).

****

Since $f_x(x,y) = 4x - y$ and $f_y(x,y) = 8y - x$, the gradient of f is

$$\nabla f(x,y) = (4x-y)\,i + (8y-x)\,j$$

and

$$\nabla f(2,1) = 7i + 6j = \langle 7, 6 \rangle$$

The unit vector, u, in the direction $\pi/3$ is

$$u = \tfrac{1}{2}i + \tfrac{\sqrt{3}}{2}j = \langle \tfrac{1}{2}, \tfrac{\sqrt{3}}{2} \rangle$$

The rate of change of $f(x,y)$ in the direction $\pi/3$ at (2,1) is the directional derivative at (2,1) in the direction $\pi/3$. This is found by forming the dot-product of u and the gradient.

$$\therefore D_u f(2,1) = u \cdot \nabla f(2,1) = \langle 7, 6 \rangle \cdot \langle \tfrac{1}{2}, \tfrac{\sqrt{3}}{2} \rangle$$

$$= \tfrac{7}{2} + 3\sqrt{3} = \frac{7 + 6\sqrt{3}}{2}$$

━━**3-67**

The surface of a certain lake is represented by a region in the xy-plane such that the depth under the point corresponding to (x,y) is:

$$f(x,y) = 300 - 2x^2 - 3y^2$$

Zeke the dog is at the point (3,2).

a) In what direction should Zeke swim in order for the depth to decrease most rapidly?

b) In what direction would the depth remain the same?

****

a) Recall that the gradient

$$\vec{\nabla} f(x,y) = \frac{\partial f}{\partial x}\,\vec{\imath} + \frac{\partial f}{\partial y}\,\vec{\jmath} = -4x\,\vec{\imath} - 6y\,\vec{\jmath}$$

gives the direction of most rapid increase and that $-\vec{\nabla} f$ gives the direction of most rapid decrease. So Zeke should swim in the direction $\quad -\vec{\nabla} f\,(3,2) = \underline{\underline{\vec{\imath} + \vec{\jmath}}}.$

b) Suppose that $\vec{d}$ is in the direction of constant depth. Then $\vec{\nabla} f \cdot \vec{d} = 0$ since the gradient is perpendicular to a level curve. So $(-\vec{\imath} - \vec{\jmath}) \cdot (d_1\vec{\imath} + d_2\vec{\jmath}) = 0 \Rightarrow d_1 = -d_2$

Thus $\underline{\vec{\imath} - \vec{\jmath}}$ and $\underline{-\vec{\imath} + \vec{\jmath}}$ are directions in which the depth would remain constant.

3-68 ■■■

Suppose that $z = x-y$, $x = 4(t^3 -1)$ and $y = \ln t$. Find $\frac{dz}{dt}$.

****

It is convenient to think of the function z diagramatically as follows:

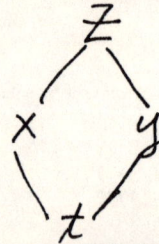

$$
\begin{array}{c}
z \\
\diagup \quad \diagdown \\
x \qquad y \\
\diagdown \quad \diagup \\
t
\end{array}
$$

Hence $\quad \dfrac{dz}{dt} = \left(\dfrac{\partial z}{\partial x}\right)\left(\dfrac{dx}{dt}\right) + \left(\dfrac{\partial z}{\partial y}\right)\left(\dfrac{dy}{dt}\right)$

Now $\dfrac{dx}{dt} = 12t^2$ and $\dfrac{dy}{dt} = \dfrac{1}{t}$

So $\quad \dfrac{dz}{dt} = (1)(12t^2) + (-1)\left(\dfrac{1}{t}\right)$

$$= 12t^2 - \dfrac{1}{t}$$

3-69

Find the function with $(3x^2 + 6y)\mathbf{i} + (3y^2 + 6x)\mathbf{j}$ as its gradient.

****

$$\frac{\delta}{\delta y}(3x^2 + 6y) = 6 \qquad \frac{\delta}{\delta x}(3y^2 + 6x) = 6$$

THUS, THE EXPRESSION IS A GRADIENT

$$\frac{\delta}{\delta x} f(x,y) = 3x^2 + 6y \qquad \frac{\delta}{\delta y} f(x,y) = 3y^2 + 6x$$

$$f(x,y) = x^3 + 6xy + \alpha(y) \qquad f(x,y) = y^3 + 6xy + \beta(x)$$

$$x^3 + 6xy + \alpha(y) = y^3 + 6xy + \beta(x)$$

$$x^3 - \beta(x) = y^3 - \alpha(y)$$

THIS IS POSSIBLE ONLY IF $\beta(x) = x^3 + C$

THUS, $f(x,y) = y^3 + 6xy + x^3 + C$

3-70

Given $f(x) = \ln(x^2 + y^2 + z^2)$.

Find $\vec{\nabla} \cdot \vec{\nabla} f$.

****

$$\vec{\nabla} f = \frac{2x}{x^2+y^2+z^2}\,\vec{i} + \frac{2y}{x^2+y^2+z^2}\,\vec{j} + \frac{2z}{x^2+y^2+z^2}\,\vec{k}$$

$$\vec{\nabla} \cdot \vec{\nabla} f = \frac{2(x^2+y^2+z^2)-2x(2x)}{(x^2+y^2+z^2)^2} + \frac{2(x^2+y^2+z^2)-(2y)(2y)}{(x^2+y^2+z^2)^2}$$

$$+ \frac{2(x^2+y^2+z^2)-(2z)(2z)}{(x^2+y^2+z^2)^2}$$

$$= \frac{2(x^2+y^2+z^2)}{(x^2+y^2+z^2)^2} = \boxed{\frac{2}{x^2+y^2+z^2}}$$

TANGENT LINES, TANGENT PLANES AND NORMAL LINES

3-71

The two surfaces $z = \sqrt{x^2 + y^2}$ and $x^2 + 4y^2 + 4z^2 = 173$ intersect in a space curve containing the point P_0 (3, 4, 5). Find a vector tangent to this curve at this point.

****

We want a vector in both tangent planes, or perpendicular to both normal vectors. One the first surface ($z^2 = x^2 + y^2$), a normal is $\langle 2x, 2y, -2z \rangle$, or $\langle 6, 8, -10 \rangle$ at the point (3,4,5). On the second surface a normal is $\langle 2x, 8y, 8z \rangle$ ($= \langle 6, 32, 40 \rangle$). Taking the vector cross product,

$$\langle 6, 8, -10 \rangle \times \langle 6, 32, 40 \rangle$$

$$= \langle 320 + 320, -60 - 240, 192 - 48 \rangle$$

$$= \langle 640, -300, 144 \rangle \text{ is tangent to}$$

the curve of intersection.

3-72 ■■

Find the equation of the plane tangent to the surface xyz = 2 at (1,-1,-2).

****

$$\phi = XYZ - 2$$

$$\vec{\nabla}\phi = (YZ, XZ, XY)\Big|_{(1,-1,-2)} = (2,-2,-1)$$

$$2(X-1) - 2(Y+1) - 1(Z+2) = 0$$

$$2X - 2 - 2Y - 2 - Z - 2 = 0 \rightarrow \underline{\underline{2X - 2Y - Z = 6}}$$

3-73 ■■

Find an equation of the plane tangent to the surface $f(x,y) = x^2 y - xy + 2$ at the point P(2,1,4).

****

We need a point the plane goes through, and a normal vector to the plane.   A point is P(2,1,4).

If $f(x,y) = z = x^2 y - xy + 2$, define

$F(x,y,z) = z - x^2 y + xy - 2 = 0$. Then $\vec{\nabla F}$ is

normal to the surface. $\vec{\nabla F}(x,y,z) = (-2xy+y)\vec{i}$

$+ (-x^2+x)\vec{j} + (1)\vec{k}$,   and $\vec{\nabla F}(2,1,4) = (-2\cdot2\cdot1+1)\vec{i}$

$+ (-2^2+2)\vec{j} + (1)\vec{k} = (-4+1)\vec{i} + (-4+2)\vec{j} + \vec{k} = -3\vec{i} - 2\vec{j} + \vec{k}$

Therefore an equation of the desired plane is

$$-3(x-2) - 2(y-1) + 1(z-4) = 0.$$

3-74

Consider the equation $x^2 + y^2 + z^2 = 49$.

1. Sketch this surface.

2. Find an equation of the tangent plane to the surface at the point (6, 2, 3).

3. Find an equation (or equations) of the line perpendicular to the tangent plane at the point (6, 2, 3).

****

1. The surface is a sphere with radius 7 and center at the origin.

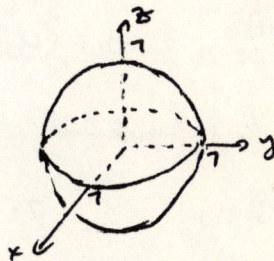

2. Let $f(x,y,z) = x^2 + y^2 + z^2 - 49$ and find $\vec{\nabla}f$, the gradient of f.

$$\vec{\nabla}f = \frac{\partial f}{\partial x}\vec{\imath} + \frac{\partial f}{\partial y}\vec{\jmath} + \frac{\partial f}{\partial z}\vec{k} = 2x\vec{\imath} + 2y\vec{\jmath} + 2z\vec{k}$$

At (6,2,3), $\vec{\nabla}f = 12\vec{\imath} + 4\vec{\jmath} + 6\vec{k}$. This is a vector perpendicular to the required plane. An equation of the plane is

$$12(x-6) + 4(y-2) + 6(z-3) = 0,$$
$$12x + 4y + 6z = 98,$$
$$6x + 2y + 3z = 49.$$

3. $\vec{\nabla}f = 12\vec{\imath} + 4\vec{\jmath} + 6\vec{k} = 2(6\vec{\imath} + 2\vec{\jmath} + 3\vec{k})$ is a vector parallel to the required line. A vector equation of the line is

$$\vec{P} = <6,2,3> + t<6,2,3>.$$

Parametric equations of the line are

$$x = 6t+6, \quad y = 2t+2, \quad z = 3t+3,$$

and symmetric equations are

$$\frac{x-6}{12} = \frac{y-2}{4} = \frac{z-3}{6}, \quad \text{or} \quad \frac{x-6}{6} = \frac{y-2}{2} = \frac{z-3}{3}.$$

Notice the gradient vector is parallel to the position vector of the point (6,2,3). This is a 3-dimensional analogy of plane geometry theorem that the tangent line to a circle is perpendicular to the radius at the point of tangency.

3-75 ■■

Let S be the surface $x^2y + 4xz^3 - yz = 0$. An equation of the tangent plane to S at $(1,2,-1)$ is (a) $y + 5z = -3$ (b) $2x - 3y + z = 3$ (c) $x - 3z = 4$ (d) $x - y + z = 1$ (e) $2x + y + 5z = -1$.

****

A normal vector for the tangent plane is $\vec{\nabla}F(1,2,-1)$ where

$F = x^2y + 4xz^3 - yz$. $\vec{\nabla}F = \langle 2xy + 4z^3, x^2-z, 12xz^2-y \rangle$,

$\vec{\nabla}F(1,2,-1) = \langle 0,2,10 \rangle = 2\langle 0,1,5 \rangle$. So the tangent plane

has equation $y + 5z = d$. Putting in $(1,2,-1)$, $d = $

$2-5 = -3$. Thus $y+5z = -3$ is the required equation.

3-76 ■■

For the surface given by $2x^2 - 3y^2 + xz - 6 = 0$, find:

a. symmetric equations of the normal line at $(2,-2,5)$.
b. the equation of the tangent plane at $(2,-2,5)$.

****

This is already in the form $F(x,y,z)=0$, so we can immediately calculate the gradient: $(4x+z)i - 6yj + xk$

At $(2,-2,5)$ this is $13i + 12j + 2k$

a. The components are the direction numbers of the normal line.

Thus: $\dfrac{x-2}{13} = \dfrac{y+2}{12} = \dfrac{z-5}{2}$

b. The components are the coefficients of the variables in the equation of the tangent plane. Thus:

$$13(x-2) + 12(y+2) + 2(z-5) = 0$$

which, to put it in best form, simplifies to

$$13x + 12y + 2z - 12 = 0$$

■■■**3-77**

Find symmetric equations of the tangent line to the curve of intersection of the surfaces

$$4x^2 - y^2 + z^2 = -11 \text{ and } x^2 + y^2 + z^3 = 18$$

at the point $(1,4,-1)$.

****

Let $f_1(x,y,z) = 4x^2 - y^2 + z^2 + 11$ and
$f_2(x,y,z) = x^2 + y^2 + z^2 - 18$.

Finding the gradient of each produces
$\nabla f_1(x,y,z) = 8x\,i - 2y\,j + 2z\,k$ and
$\nabla f_2(x,y,z) = 2x\,i + 2y\,j + 2z\,k$.

$\nabla f_1(1,4,-1) = 8i - 8j - 2k$ is a normal vector at
$(1,4,-1)$ to the surface $f_1(x,y,z) = 0$ and

$\nabla f_2(1,4,-1) = 2i + 8j - 2k$ is a normal vector at
$(1,4,-1)$ to the surface $f_2(x,y,z) = 0$.

The components of the cross-product of these two normal vectors are a set of direction numbers of the tangent line.

$$\nabla f_1(1,4,-1) \times \nabla f_2(1,4,-1) = \begin{vmatrix} i & j & k \\ 8 & -8 & -2 \\ 2 & 8 & -2 \end{vmatrix}$$

$$= \begin{vmatrix} -8 & -2 \\ 8 & -2 \end{vmatrix} i - \begin{vmatrix} 8 & -2 \\ 2 & -2 \end{vmatrix} j + \begin{vmatrix} 8 & -8 \\ 2 & 8 \end{vmatrix} k$$

$$= 32i + 12j + 80k = 4(8i + 3j + 20k)$$

Thus $\{8,3,20\}$ are direction numbers for the tangent line.

Then symmetric equations of the tangent line are

$$\frac{x-1}{8} = \frac{y-4}{3} = \frac{z+1}{20}$$

3-78 ■■

Show that each normal line of $(x-h)^2 + (y-k)^2 + (z-t)^2 = r^2$ contains the point (h,k,t).

****

Let the point $P(x_0, y_0, z_0)$ be on the surface of the sphere and $F(x,y,z) = (x-h)^2 + (y-k)^2 + (z-t)^2 - r^2$.

In order to determine a normal line of the sphere one needs a direction and point. The desired direction is $\langle F_x, F_y, F_z \rangle \big|_P$ $= \langle 2(x-h), 2(y-k), 2(z-t) \rangle \big|_P = \langle 2(x_0-h), 2(y_0-k), 2(z_0-t) \rangle$.

With this direction and point P we have the general form of a normal line

$$\begin{cases} x = x_0 + 2(x_0 - h)V \\ y = y_0 + 2(y_0 - k)V \\ z = z_0 + 2(z_0 - t)V. \end{cases}$$

Note that V is an element of the reals and in particular $V = -\frac{1}{2} \Rightarrow (h,k,t)$ is contained on the normal line. Therefore each normal of a sphere contains its center.

■■■ **3-79**

Find an equation of the tangent plane to the surface

$$4x^2 - y^2 + 3z^2 = 10$$

at the point $(2,-3,1)$.

****

To find an equation of the tangent plane we need a point on the surface and the direction normal to the surface at that point.

The normal direction can be found by evaluating the three-dimensional gradient

$$\vec{\nabla} f(x,y,z) = \frac{\partial f}{\partial x} \vec{i} + \frac{\partial f}{\partial y} \vec{j} + \frac{\partial f}{\partial z} \vec{k}$$

$$= (8x)\vec{i} - (2y)\vec{j} + (6z)\vec{k}$$

So $\vec{\nabla} f(2,-3,1) = 16\vec{i} + 6\vec{j} + 6\vec{k}$

Thus, an equation of the tangent plane is

$$16(x-2) + 6(y+3) + 6(z-1) = 0$$

or

$$\underline{16x + 6y + 6z = 20}$$

3-80 ■■

Show that the sum of the squares of the x, y, and z intercepts of the tangent plane of $x^{2/3} + y^{2/3} + z^{2/3} = a^{2/3}$ is a^2.

****

First we must find the tangent plane of the given surface. Let $F(x,y,z) = x^{2/3} + y^{2/3} + z^{2/3} - a^{2/3}$ and let the point $P(x_0, y_0, z_0)$ be on the surface. The normal of the tangent plane at P is $\langle F_x, F_y, F_z \rangle \big|_P = \langle \frac{2}{3} x^{-1/3}, \frac{2}{3} y^{-1/3}, \frac{2}{3} z^{-1/3} \rangle \big|_P = \frac{2}{3} \langle x_0^{-1/3}, y_0^{-1/3}, z_0^{-1/3} \rangle$. Hence the tangent plane is $\langle x_0^{-1/3}, y_0^{-1/3}, z_0^{-1/3} \rangle \cdot \langle x - x_0, y - y_0, z - z_0 \rangle = 0 \Rightarrow x_0^{-1/3} x - x_0^{2/3} + y_0^{-1/3} y - y_0^{2/3} + z_0^{-1/3} z - z_0^{2/3} = 0 \Rightarrow x_0^{-1/3} x + y_0^{-1/3} y + z_0^{-1/3} z - (x_0^{2/3} + y_0^{2/3} + z_0^{2/3}) = 0 \Rightarrow x_0^{-1/3} x + y_0^{-1/3} y + z_0^{-1/3} z - a^{2/3} = 0$. The intercepts are $a^{2/3} x_0^{1/3}$, $a^{2/3} y_0^{1/3}$, and $a^{2/3} z_0^{1/3}$. Now the sum of the squares of the intercepts is $(a^{2/3} x_0^{1/3})^2 + (a^{2/3} y_0^{1/3})^2 + (a^{2/3} z_0^{1/3})^2 = a^{4/3} (x_0^{2/3} + y_0^{2/3} + z_0^{2/3}) = a^{4/3} (a^{2/3}) = a^{6/3} = a^2$.

■■■**3-81**

Find the equation of the tangent plane to the surface z= f(x, y)= $x^3 y^4$ at the point (-1, 2, -16).

****

Let (x, y, z) be any point in the tangent plane. Then the vector $(x -(-1), y-2, z-(-16))$

$= (x+1, y-2, z+16)$ is a vector through $(-1, 2, -16)$ that lies in the tangent plane.

Now $x^3 y^4 - z = 0$ is a level surface of the surface $w = x^3 y^4 - z$; hence, $\vec{\nabla w}$ is perpendicular to $z = x^3 y^4$. and hence $\vec{\nabla w}$ is perpendicular to $(x+1, y-2, z+16)$.

Now $\vec{\nabla w}_{(x,y,z)} = (3x^2 y^4, 4x^3 y^3, -1)$ so

$\vec{\nabla w}(-1, 2, -16) = (48, -32, -1)$

Thefore the equation of the tangent plane is given by:

$(48, -32, -1) \cdot (x+1, y-2, z+16) = 0$

$48(x+1) - 32(y-2) - (z+16) = 0$

or

$48x - 32y - z + 96 = 0$

3-82 ■■

Show that the two spheres given by the equations

$$x^2 + y^2 + z^2 = 9$$

$$(x - 2)^2 + y^2 + z^2 = 1$$

are tangent at the point (3,0,0).

****

First, it must be verified that the point $(3,0,0)$ is a point on both spheres :

$$3^2 + 0^2 + 0^2 = 9 \quad \checkmark$$
$$(3-2)^2 + 0^2 + 0^2 = 1 \quad \checkmark$$

thus $(3,0,0)$ lies on both spheres.

To show that the 2 spheres are tangent at $(3,0,0)$, it is sufficient to show that they have the same tangent plane :

The normal vector to $x^2 + y^2 + z^2 = 9$ is given by

$$\vec{N_1} = 2x\,\vec{i} + 2y\,\vec{j} + 2z\,\vec{k},$$

evaluated at $(3,0,0)$ the vector is $6\vec{i}$

thus the equation of the tangent plane is $6x = 18$

OR $x = 3$

The normal vector to $(x-2)^2 + y^2 + z^2 = 1$ is given by

$$\vec{N_2} = 2(x-2)\,\vec{i} + 2y\,\vec{j} + 2z\,\vec{k}$$

evaluated at $(3,0,0)$ the vector is $2\vec{i}$

thus the equation of the tangent plane

is $\quad 2x = 6 \quad$ or $\quad x = 3$.

Since $(3, 0, 0)$ is on both spheres and the plane

$x = 3$ is tangent to both spheres at $(3, 0, 0)$,

the 2 spheres are tangent

■■**3-83**

Given $f(x,y) = xe^{-2y}$, at the point $(4,0)$,
     a. find the equation of the tangent plane to the graph of f,
     b. find a normal vector to the graph of f.

****

THE PARTIAL DERIVATIVES ARE
$$\frac{\partial f}{\partial x}(x,y) = e^{-2y}$$
$$\frac{\partial f}{\partial y}(x,y) = -2xe^{-2y}$$
AT $(4,0)$
$$\frac{\partial f}{\partial x}(4,0) = e^{0} = 1$$
$$\frac{\partial f}{\partial y}(4,0) = -2 \cdot 4 \cdot e^{0} = -8.$$

a) THE EQUATION OF THE TANGENT PLANE IS
$$z = f(4,0) + 1(x-4) + (-8)(y-0)$$
$$z = 4 \cdot e^{0} + x - 4 - 8y$$
$$z = x - 8y.$$

b) TWO VECTORS NORMAL TO THE GRAPH OF
f AT $(4,0)$ ARE
$$[1, -8, -1] \quad \text{AND} \quad [-1, 8, 1].$$

3-84 ■■■

Consider the surface given by $z = xy^3 - x^2y$.

(a) Find an equation for the tangent plane to the surface at the point $(3,2,6)$.

(b) Find parametric equations for the normal line to the surface at the point $(3,2,6)$.

****

(a) Let $f(x,y) = xy^3 - x^2y$.

$$f_x(x,y) = y^3 - 2xy, \quad f_x(3,2) = -4$$

$$f_y(x,y) = 3xy^2 - x^2, \quad f_y(3,2) = 27$$

$\therefore$ $\langle -4, 27, -1 \rangle$ is normal to surface

Equation: $-4(x-3) + 27(y-2) - (z-6) = 0$

or, $4x - 27y + z = -36$

(b) $x = 3 - 4t, \quad y = 2 + 27t, \quad z = 6 - t$

3-85 ■■■

Find the equation of the tangent plane to the surface $xy^2 - 3xz = -3$ at the point $(-3,-2,1)$.

****

$$f(x,y,z) = xy^2 - 3xz + 3$$

$$(\text{grad } f) = (y^2 - 3z, 2xy, -3x)$$

$$\left(\operatorname{grad} f\right)_{(-3,-2,1)} = (1, 12, 9). \quad \text{THIS IS } \underline{N} \text{ THE}$$

NORMAL VECTOR TO THE PLANE.

$$\underline{x}_0 = (-3, -2, 1)$$

$$0 = \underline{N} \cdot (\underline{x} - \underline{x}_0) = (1, 12, 9) \cdot (x+3, y+2, z-1)$$

So $\quad 0 = x + 12y + 9z + 18.$

■■**3-86**

Let C be the curve of intersection of $x^2 + y^2 - z = 8$ and $x - y^2 + z^2 = -2$.
Parametric equations of the tangent line to C at the point $(2, -2, 0)$ are
(a) $x = 4 + 4t$, $y = -1 - 4t$, $z = 20 - t$ (b) $x = 2 + 4t$, $y = -2 - t$, $z = 20t$
(c) $x = 2 + 4t$, $y = -2 - 4t$, $z = -t$ (d) $x = 2 + 3t$, $y = -2 + t$, $z = -5t$
(e) $x = 2 + 2t$, $y = -2 - -2t$, $z = t$.

****

Let $F = x^2 + y^2 - z - 8$, $\quad G = x - y^2 + z^2 + 2.$

$\vec{\nabla F} = \langle 2x, 2y, -1 \rangle$, $\quad \vec{\nabla G} = \langle 1, -2y, 2z \rangle.$

$\vec{N_1} = \vec{\nabla F}(2, -2, 0) = \langle 4, -4, -1 \rangle$, $\vec{N_2} = \vec{\nabla G}(2, -2, 0) = \langle 1, 4, 0 \rangle.$

Now the tangent line to C is the intersection of the 2

tangent planes to the 2 surfaces which have normal

vectors $\vec{N_1}, \vec{N_2}$. Hence $\vec{N_1} \times \vec{N_2}$ is in the direction of

the tangent line of C. $\vec{N_1} \times \vec{N_2} = \begin{vmatrix} \vec{i} & \vec{j} & \vec{k} \\ 4 & -4 & -1 \\ 1 & 4 & 0 \end{vmatrix} = \langle 4, -1, 20 \rangle.$

So, $x = 2 + 4t$, $y = -2 - t$, $z = 20t$ is the required line.

EXTREME VALUES

3-87 ■■■

Find the critical points of the surface
$$f(x,y) = 3x^3 + y^2 - 9x + 4y$$
and classify these points into maximums, minimums, and saddle points.

****

$f_x = 9x^2 - 9 = 9(x^2 - 1)$ $f_y = 2y + 4 = 2(y + 2)$

Critical points occur when $f_x = f_y = 0$

$\quad x^2 - 1 = 0$, so $x = \pm 1$ $y + 2 = 0$, so $y = -2$

So, critical points occur at $(1, -2)$ and $(-1, -2)$

$D = f_{xx} f_{yy} - f_{xy}^2$

$f_{xx} = 9 \cdot 2x = 18x$ $f_{yy} = 2$ $f_{xy} = 0$

Hence $D = 36x$

at $x = 1$ $D = 36 > 0$ and $f_{xx} = 18 > 0$

$\qquad$ so $(1, -2)$ is a relative minimum

at $x = -1$ $D = -36 < 0$ so $(-1, -2)$ is a saddle point.

■■■ ■■■■■■■■■■■■■■■■■■■■■■■■■■■■■■■■■■■ **3-88**

Find all relative maximums, relative minimums, and saddle points of
$f(x,y) = x^3 + 3xy^2 - 3x^2 - 3y^2 + 4$.

****

$f_x = 3x^2 + 3y^2 - 6x$, $f_y = 6xy - 6y$. Setting $f_x = 0$ and $f_y = 0$

gives (a) $x^2 + y^2 - 2x = 0$.   (b) is true for $y = 0$ or $x = 1$.
(b) $y(x-1) = 0$

For $y = 0$, (a) becomes $x^2 - 2x = 0$, $x(x-2) = 0$, $x = 0, x = 2$.

For $x = 1$, (a) gives $1 + y^2 - 2 = 0$, $y^2 = 1$, $y = 1$, $y = -1$.

So the critical points are $(0,0)$, $(2,0)$, $(1,1)$, $(1,-1)$.

To test the critical points, $f_{xx} = 6x - 6$, $f_{yy} = 6x - 6$, $f_{xy} = 6y$,

$\varphi \equiv f_{xx} \cdot f_{yy} - f_{xy}^2 = (6x-6)^2 - 36y^2 = 36[(x-1)^2 - y^2]$.

$\varphi(0,0) = 36 > 0$ and $f_{xx}(0,0) = -6 < 0 \Rightarrow f(0,0) = 4$ is a rel. max.

$\varphi(2,0) = 36 > 0$ and $f_{xx}(2,0) = 6 > 0 \Rightarrow f(2,0) = 0$ is a rel. min.

$\varphi(1,1) = \varphi(1,-1) = -36 < 0 \Rightarrow (1,1,2)$ and $(1,-1,2)$ are saddle points.

3-89 ■■■

The function z = xy + (x + y)(120 - x - y) has a maximum. Find the values of x and y at which it occurs.

****

Find $\frac{dz}{dx}$ and $\frac{dz}{dy}$, set them equal to 0, and solve the resulting equations simultaneously.

$$\frac{dz}{dx} = y + (x+y)(-1) + (120-x-y)(1) = 120 - 2x - y$$

$$\frac{dz}{dy} = x + (x+y)(-1) + (120-x-y)(1) = 120 - x - 2y$$

$$\frac{dz}{dx} = 0 \implies 2x + y = 120, \quad 4x + 2y = 240$$

$$\frac{dz}{dy} = 0 \implies x + 2y = 120, \quad \underline{x + 2y = 120}$$

$$3x \qquad = 120, \quad x = 40$$

Substitute x = 40 into 2x + y = 120 to find y = 40. Both partials exist everywhere. Hence, if z attains a maximum value, it must do so at a point where $dz/dx = 0 = dz/dy$. We have found the only such point, x = 40 = y. Assuming the problem to be correctly stated (that is, assuming a maximum exists), it must occur at x = 40 = y. You can check that this point actually yields a maximum using a second derivative test if you want.

■■**3-90**

Locate and classify all relative maxima, relative minima, and saddle points of $g(x,y) = x^3 - 3xy - y^3$.

****

$$g_x(x,y) = 3x^2 - 3y = 0 \implies y = x^2$$

$$g_y(x,y) = -3x - 3y^2 = 0 \implies x = -y^2$$

$$x = -y^2 = -(x^2)^2 = -x^4$$

$$x^4 + x = 0, \quad x(x^3 + 1) = 0, \quad x = 0, -1$$

$$x = 0 \text{ gives } y = 0, \quad x = -1 \text{ gives } y = 1$$

$\therefore$ critical points are $(0,0)$ and $(-1,1)$

$$g_{xx}(x,y) = 6x, \quad g_{xy}(x,y) = -3, \quad g_{yy}(x,y) = -6y$$

$\underline{(0,0)}:$ $D = g_{xx} g_{yy} - (g_{xy})^2 = 0 - 9 < 0$

$\qquad \therefore$ saddle point

$\underline{(-1,1)}:$ $D = g_{xx} g_{yy} - (g_{xy})^2 = (-6)(-6) - 9 > 0$

$\qquad \therefore$ since $g_{xx} < 0$, relative maximum

3-91 ■■■

Examine for maxima and minima: $z = 3x^2 - 6xy + 6y^2 + 3x - 9y + 10.$

****

$$\frac{\partial z}{\partial x} = 6x - 6y + 3 = 0 \quad , \quad y = \frac{6x+3}{6}$$

$$\frac{\partial z}{\partial y} = -6x + 12y - 9 = 0 \quad , \quad -6x + 12\left(\frac{6x+3}{6}\right) - 9 = 0$$

$$-6x + 12x + 6 - 9 = 0 \quad , \quad 6x = 3 \quad , \quad x = \tfrac{1}{2}$$

$$y = \frac{6x+3}{6} = \frac{6\left(\tfrac{1}{2}\right)+3}{6} = 1$$

$$\frac{\partial^2 z}{\partial x^2} = 6 \quad , \quad \frac{\partial^2 z}{\partial y^2} = 12 \quad , \frac{\partial^2 z}{\partial x \partial y} = -6$$

$$\Delta = \frac{\partial^2 z}{\partial x^2} \cdot \frac{\partial^2 z}{\partial y^2} - \left(\frac{\partial^2 z}{\partial x \partial y}\right)^2 \quad \text{at} \quad x = \tfrac{1}{2}, \, y = 1$$

$$= 6 \cdot 12 - (-6)^2 = 72 - 36 = 36$$

If $\Delta > 0$ and $\frac{\partial^2 z}{\partial x^2} > 0$, the point is a local minimum. Thus, $\left(\tfrac{1}{2}, 1\right)$ is a local minimum.

━━ 3-92

Determine the relative extrema of f if there are any where

$$f(x,y) = x^2 + y^2 - xy + x - 5y$$

****

To apply the second-derivative test, find

$f_x(x,y) = 2x - y + 1$ $\qquad f_{xx}(x,y) = 2$

$f_y(x,y) = 2y - x - 5$ $\qquad f_{yy}(x,y) = 2$ $\qquad f_{xy}(x,y) = -1$

Set $f_x(x,y) = 0$ and $f_y(x,y) = 0$ and solve.

$$\begin{array}{lll} 2x - y + 1 = 0 & 2x - y = -1 & 2x - y = -1 \\ 2y - x - 5 = 0 \;\; \text{or} & -x + 2y = 5 \;\; \text{or} & \underline{-2x + 4y = 10} \\ & & 3y = 9 \\ & & y = 3 \end{array}$$

Substituting $y = 3$ into

$\qquad\qquad 2x - y + 1 = 0$ gives $x = 1$

Therefore $(1,3)$ is the critical point of f and

$f_{xx}(1,3) = 2$, $f_{yy}(1,3) = 2$ and $f_{xy}(1,3) = -1$

Since $f_{xx}(1,3) \cdot f_{yy}(1,3) - f_{xy}^2(1,3) = (2)(2) - (-1)^2$

$$= 3 > 0,$$

f has a relative minimum value at $(1,3)$. That value is

$$f(1,3) = 1 + 9 - 3 + 1 - 15 = -7$$

3-93 ■■■

Given $f(x,y) = -x^4+4xy-2y^2+1$, use the second derivative test to find all local extreme points and saddle points.

****

THE PARTIAL DERIVATIVES ARE

$$f_x = -4x^3+4y = -4(x^3-y)$$
$$f_y = 4x-4y = 4(x-y).$$

CRITICAL POINTS:

$$f_y = 0 \implies y = x$$
$$f_x = 0 \implies x^3-y = x^3-x = 0$$
$$x(x+1)(x-1) = 0$$
$$x = -1, 0, 1$$

$$(-1,-1), (0,0), (1,1)$$

SECOND DERIVATIVE TEST:

$$f_{xx} = -12x^2$$
$$f_{yy} = -4$$
$$f_{xy} = 4$$

POINT	f_{xx}	f_{yy}	f_{xy}	$f_{xy}^2 - f_{xx}f_{yy}$	CONCLUSION
$(-1,-1)$	-12	-4	4	-32	LOCAL MAX.
$(0,0)$	0	-4	4	16	SADDLE PT.
$(1,1)$	-12	-4	4	-32	LOCAL MAX.

━━━━━━━━━━━━━━━━━━━━━━━━━━━━━━━━━━━━━━━ **3-94**

Find all local maxima, local minima, and saddle points for the function
$f(x,y) = 2x^3 + 4y^3 + 3x^2 - 12x - 192y + 5.$

****

$$f_x = 6x^2 + 6x - 12 = 6(x^2 + x - 2) = 6(x-2)(x+1)$$

$$f_y = (12y^2 - 192) = 12(y^2 - 16) = 12(y+4)(y-4)$$

So critical pts. are. $(2,4)$; $(2,-4)$;
$\qquad\qquad\qquad (-1,4)$; $(-1,-4)$

$$f_{xx} = 6(2x+1)$$

$$f_{yy} = 24y.$$

$$f_{xy} = f_{yx} = 0$$

$$f_{xx}f_{yy} - (f_{xy})^2 = 144y(2x-1) = \Delta(x,y).$$

$\Delta(2,4) = 144(4)(3) > 0$ $\Big\}$ $\Rightarrow (2,4)$ is local.
$\quad$ and $f_{xx} > 0$ at $(2,4)$ $\qquad\qquad$ min.

$\Delta(2,-4) = 144(-4)(3) < 0 \Rightarrow (2,-4)$ is a
$\qquad\qquad\qquad\qquad\qquad\qquad$ saddle point

$\Delta(-1,4) = 144(4)(-1) < 0 \Rightarrow (-1,4)$ is a
$\qquad\qquad\qquad\qquad\qquad\qquad$ saddle point

$\Delta(-1,-4) = 144(-4)(-1) > 0$ $\Big\}$ $\Rightarrow (-1,-4)$ is a
$\quad f_{xx} < 0$ at $(-1,-4)$ $\qquad$ local max.

3-95 ■■

Find all critical points for $f(x,y) = x^3 + y^3 - 3xy + 5$, and determine for each one if it is a relative minimum, relative maximum, or saddle point.

****

$$\frac{\partial f}{\partial x} = 3x^2 - 3y , \qquad \frac{\partial^2 f}{\partial x^2} = 6x$$

$$\frac{\partial f}{\partial y} = 3y^2 - 3x , \qquad \frac{\partial^2 f}{\partial y^2} = 6y \qquad \frac{\partial^2 f}{\partial x \partial y} = -3$$

$$\frac{\partial f}{\partial x} = 0 \Rightarrow y = x^2 , \qquad \frac{\partial f}{\partial y} = 0 \Rightarrow x = y^2$$

But $y = x^2$ and $x = y^2$ intersect at $(0,0)$ and at $(1,1)$, so these are the critical points of f.

Let $D = \left(\frac{\partial^2 f}{\partial x^2}\right)\left(\frac{\partial^2 f}{\partial y^2}\right) - \left(\frac{\partial^2 f}{\partial x \partial y}\right)^2$

$$= (6x)(6y) - 9 = 9(4xy - 1)$$

$D(0,0) = -9 \Rightarrow (0,0)$ is a <u>saddle point</u> of f.

$D(1,1) = 27 > 0 , \quad \frac{\partial^2 f}{\partial x^2}(1,1) = 6 > 0 \Rightarrow$

$(1,1)$ is a point where f has a <u>local minimum</u>.

■■■**3-96**

Find all critical values of $f(x,y) = 4xy - x^4 - y^4 + \frac{1}{16}$ and classify each one.

****

$$\frac{\delta f}{\delta x} = 4y - 4x^3 \qquad\qquad \frac{\delta f}{\delta y} = 4x - 4y^3$$

$$4y - 4x^3 = 0 \qquad\qquad 4x - 4y^3 = 0$$

$$y = x^3 \qquad\qquad\qquad x = y^3$$

$$x = (x^3)^3 = x^9$$

$$x - x^9 = 0$$

$$x(x-1)(x+1)(x^2+1)(x^4+1) = 0$$

$$x = 0, \pm 1$$

CRITICAL VALUES ARE $(0,0)$, $(1,1)$, AND $(-1,-1)$

$$\frac{\delta^2 f}{\delta x^2} = -12x^2 \qquad\qquad \frac{\delta^2 f}{\delta y^2} = -12y^2$$

$$\frac{\delta^2 f}{\delta x \, \delta y} = 4$$

$$4^2 - \left[-12(0)^2\right]\cdot\left[-12(0)^2\right] = 16$$

THUS SADDLE POINT AT $(0,0)$

$$4^2 - \left[-12(1)^2\right]\cdot\left[-12(1)^2\right] = -128$$

THUS, RELATIVE MAXIMUM AT $(1,1)$

$$4^2 - \left[-12(-1)^2\right]\cdot\left[-12(-1)^2\right] = -128$$

THUS, RELATIVE MAXIMUM AT $(-1,-1)$

3-97 ■■

Examine the function $f(x,y) = x^2 + y^2 - 4x + 6y + 25$ for maximum or minimum points. At such points give the maximum or minimum function values.

****

Holding y constant we compute $\frac{\partial f}{\partial x} = 2x - 4$.

Holding x constant we compute $\frac{\partial f}{\partial y} = 2y + 6$. setting

$\frac{\partial f}{\partial x} = 0$ AND $\frac{\partial f}{\partial y} = 0$ yields $x_c = 2$ AND $y_c = -3$. Also,

$f(2, -3) = 4 + 9 - 8 - 18 + 25 = 12$. Now use the second derivative test to decide the type of extremum if any.

Compute: $\frac{\partial^2 f}{\partial x^2} = 2$, $\frac{\partial^2 f}{\partial y^2} = 2$, $\frac{\partial^2 f}{\partial x \partial y} = 0$. Then

$$D(x,y) = \frac{\partial^2 f}{\partial x^2} \cdot \frac{\partial^2 f}{\partial y^2} - \left(\frac{\partial^2 f}{\partial x \partial y}\right)^2 = 4 - 0 = 4 > 0 \text{ foR all } x, y.$$

Hence we have a Maximum oR minimum at $x = 2, y = -3$ since $\frac{\partial^2 f}{\partial x^2} = 2 > 0$ foR all x, y we have located a

minimum point at $(2, -3, 12)$. Also 12 is a minimum function value.

3-98

Compute the minimum value of z and sketch a portion of the graph of
$$z = 3x^2 + 6x + 2y^2 - 8y$$
near its lowest point.

****

$z_x = 6x + 6 = 0 \Rightarrow x = -1. \quad z_y = 0 = 4y - 8 \Rightarrow y = 2.$

$z_{xx} = 6. \quad z_{yy} = 4. \quad z_{xy} = 0.$

$z_{xx}(-1,2) \, z_{yy}(-1,2) - [z_{xy}(-1,2)]^2 = 24 > 0, \quad z_{xx}(-1,2) = 6 > 0$

$\Rightarrow z(-1,2) = 3 - 6 + 8 - 16 = -11$ is a relative minimum.
Since the graph is an elliptic paraboloid and
concave up, $z = -11$ is the absolute minimum. As
a check: $z = 3(x^2 + 2x) + 2(y^2 - 4y)$

$\qquad = 3(x+1)^2 + 2(y-2)^2 - 11 \geq -11$

The vertex is at $(-1, 2, -11)$.

When $z = 0, \quad \dfrac{(x+1)^2}{11/3} + \dfrac{(y-2)^2}{11/2} = 1.$

$\qquad 1 < < \frac{11}{3} < 2^2 < \frac{11}{2} < 3^2$

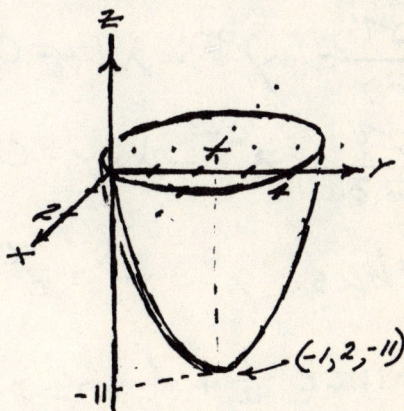

LAGRANGE MULTIPLIERS

3-99 ■■

Let a, b, and c be positive real numbers such that $\frac{1}{a} + \frac{1}{b} + \frac{1}{c} = 1$.

Find the minimum value of $\frac{x^a}{a} + \frac{y^b}{b} + \frac{z^c}{c}$

on the surface xyz = 1 in the first octant.

****

Using Lagrange multipliers,

$$\psi = \frac{x^a}{a} + \frac{y^b}{b} + \frac{z^c}{c} + \lambda(1 - xyz).$$

$$\frac{\partial \psi}{\partial x} = x^{a-1} - \lambda yz = 0, \quad x^a = \lambda xyz$$

$$\frac{\partial \psi}{\partial y} = y^{b-1} - \lambda xz = 0, \quad y^b = \lambda xyz$$

$$\frac{\partial \psi}{\partial z} = z^{c-1} - \lambda xy = 0, \quad z^c = \lambda xyz$$

Thus $x^a = y^b = z^c$, $\frac{x^a}{a} + \frac{x^a}{b} + \frac{x^a}{c} = 1$, and

since $\frac{1}{a} + \frac{1}{b} + \frac{1}{c} = 1$, $x^a = 1$, $x = 1 = y = z$, our

minimum is $\underline{1}$.

3-100

Solve completely, using Lagrange multipliers: Find the dimensions of a box with volume = 1000 which minimize the total length of the 12 edges.

****

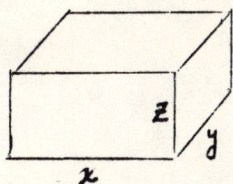

The function to be minimized is the total length of the 12 edges = $4x + 4y + 4z$.

The constraint on x, y, z is $xyz = 1000$.

So our problem is: minimize $f(x,y,z) = 4x + 4y + 4z$ subject to the constraint $g(x,y,z) = xyz - 1000 = 0$
For Lagrange multipliers we solve $\nabla f + \lambda \nabla g = 0$
This gives the vector equation $(4,4,4) + \lambda(yz, xz, xy) = 0$.
Together with $xyz = 1000$, this gives the system of eqs

(1) $\begin{cases} \lambda yz = -4 \\ \lambda xz = -4 \\ \lambda xy = -4 \\ xyz = 1000 \end{cases}$ Solve for λ, x, y, z.

One of many methods: let $x = \frac{1000}{yz}$ and reduce (1) to a system of 3 equations in 3 unknowns λ, y, z:

(2) $\begin{cases} \lambda yz = -4 \\ \lambda = -.004 y \\ \lambda = -.004 z \end{cases}$

Observe that the last two eqs $\Rightarrow y = z$, so (2) reduces to

(3) $\begin{cases} \lambda y^2 = -4 \\ \lambda = -.004 y \end{cases}$ $\Rightarrow$ $-.004 y^3 = -4$
$$y^3 = 1000$$
$$\boxed{\begin{array}{c} y = 10 \\ \Rightarrow z = 10 \text{ and } x = 10 \end{array}}$$

These dimensions minimize f, subject to the given constraint.

3-101 ■■■

Find the greatest product 3 numbers can have if the sum of their squares must be 48.

****

max xyz

subject $x^2 + y^2 + z^2 = 48$.

Let x, y and z be the 3 numbers. We wish to maximize $f(x,y,z) = xyz$ subject to the constraint $x^2 + y^2 + z^2 - 48 = 0$. To use Lagrange Multipliers, we identify the function $F(x,y,z,\lambda) = xyz + \lambda(x^2 + y^2 + z^2 - 48)$. Extremums exist where $\nabla F = 0$. $\nabla F = \langle yz + 2\lambda x, \ xz + 2\lambda y, \ xy + 2\lambda z, \ x^2 + y^2 + z^2 - 48 \rangle$ $\nabla F = 0$ iff Ⓐ $yz + 2\lambda x = 0$, Ⓑ $xz + 2\lambda y = 0$, Ⓒ $xy + 2\lambda z = 0$, and Ⓓ $x^2 + y^2 + z^2 - 48 = 0$ (4 equations, 4 unknowns)

Obviously we can get a better product xyz than we would get if x, y or z equals 0. So we require $x \neq 0, y \neq 0, z \neq 0$. From Ⓐ we get $\lambda = -yz/2x$. Putting this into Ⓑ gives $xz + 2y(-yz/2x) = 0$, or $x^2 = y^2$. Putting $\lambda = -yz/2x$ into Ⓒ gives $xy + 2z(-yz/2x) = 0$, or $x^2 = z^2$. Putting $x^2 = y^2$ and $x^2 = z^2$ into Ⓓ gives $x^2 + x^2 + x^2 = 48$, or $3x^2 = 48$, or $x^2 = 16$, or $x = \pm 4$. This implies (since $x^2 = y^2$ and $x^2 = z^2$) that $y = \pm 4$ and $z = \pm 4$. All 8 points (x,y,z) made up of x, y and $z = \pm 4$ are possible maximum points. By taking all 8 products xyz, we see that $x = 4$, $y = 4$ and $z = 4$ gives the greatest value (as do 3 other points), and that greatest product is 64.

3-102

Optimize $f(x,y) = x^2 + y^2 + 2$ subject to $xy = 4$.

****

To use a Lagrange multiplier an auxiliary function, F, is formed as

$$F(x,y,\lambda) = f(x,y) + \lambda g(x,y) \text{ where } g(x,y) = xy - 4$$

so $F(x,y,\lambda) = x^2 + y^2 + 2 + \lambda xy + 4\lambda$. To find the critical points, determine $F_x(x,y,\lambda)$, $F_y(x,y,\lambda)$ and $F_\lambda(x,y,\lambda)$. Set each equal to zero and solve simultaneously.

$F_x(x,y,\lambda) = 2x + \lambda y = 0$ ①

$F_y(x,y,\lambda) = 2y + \lambda x = 0$ ②

$F_\lambda(x,y,\lambda) = xy - 4 = 0 \rightarrow y = \dfrac{4}{x}$ ③

Substituting $y = \dfrac{4}{x}$ into equations ① and ② gives

$2x + \dfrac{4\lambda}{x} = 0$ and $\dfrac{8}{x} + \lambda x = 0$

$2x^2 + 4\lambda = 0$ $8 + \lambda x^2 = 0$

$\lambda = -\dfrac{x^2}{2}$ $\lambda = -\dfrac{8}{x^2}$

Since $\lambda = \lambda$, $-\dfrac{x^2}{2} = -\dfrac{8}{x^2}$ or $x^4 = 16$

$x^2 = 4 \text{ or } -4$

$x = \pm 2$ and $y = \dfrac{4}{x}$

$\therefore (2,2)$ and $(-2,-2)$ are critical points.

$f_x(x,y) = 2x$ $f_{xx}(x,y) = 2$

$f_y(x,y) = 2y$ $f_{yy}(x,y) = 2$ $f_{xy}(x,y) = 0$

Therefore by the second derivative test

$f_{xx}(2,2) \cdot f_{yy}(2,2) - f_{xy}^2(2,2) = (2)(2) - 0 = 4 > 0$ so

$f(2,2) = 10$ is a relative minimum value.

also

$f_{xx}(-2,-2) \cdot f_{yy}(-2,-2) - f_{xy}^2(-2,-2) = (2)(2) - 0 = 4 > 0$

and $f(-2,-2) = 10$ is also a relative minimum value.

3-103 ■■■

Find the point on the plane $x - 2y + z = 3$ where $x^2 + 4y^2 + 2z^2$ is minimum.

****

$$f(x,y,z) = x^2 + 4y^2 + 2z^2$$

CONSTRAINT IS. $x - 2y + z - 3 = 0$

$$F(x,y,z) = x^2 + 4y^2 + 2z^2 - \lambda(x - 2y + z - 3)$$

$F_x = 2x - \lambda.$ $F_x = 0 \iff x = \lambda/2$

$F_y = 8y + 2\lambda.$ $F_y = 0 \iff y = -\lambda/4$

$F_z = 4z - \lambda.$ $F_z = 0 \iff z = \lambda/4.$

$F_\lambda = -(x - 2y + z - 3)$ $F_\lambda = 0 \iff x - 2y + z - 3 = 0.$

So

$$\frac{\lambda}{2} + \frac{\lambda}{2} + \frac{\lambda}{4} = 3$$

$$\frac{5\lambda}{4} = 3$$

$$\lambda = 12/5$$

So $\left. \begin{array}{l} x = 4/5 \\ y = -3/5 \\ z = 3/5 \end{array} \right\}$ This is desired point

━━━━━━━━━━━━━━━━━━━━━━━━━━━━━━━━━━━━**3-104** ✓

Maximize $3x - y + 1$ on the ellipse $3x^2 + y^2 = 16$

****

LET $f(x,y) = 3x - y + 1$, $g(x,y) = 3x^2 + y^2 - 16$

$$\nabla f(x,y) = 3i - j \qquad \nabla g(x,y) = 6xi + 2yj$$

$$3i - j = \lambda(6xi + 2yj)$$

$$3 = 6x\lambda \qquad\qquad -1 = 2y\lambda$$

$$x = \frac{1}{2\lambda} \qquad\qquad y = -\frac{1}{2\lambda}$$

$$-x = y$$

$$3x^2 + (-x)^2 = 16$$
$$4x^2 = 16$$
$$x = \pm 2$$

POINTS UNDER CONSIDERATION:
$(2, -2), (-2, 2)$

$$f(2, -2) = 3(2) - (-2) + 1 = 9$$

$$f(-2, 2) = 3(-2) - 2 + 1 = -7$$

MAXIMUM VALUE IS 9

━━━━━━━━━━━━━━━━━━━━━━━━━━━━━━━━━━━━**3-105** ∿

Compute the minimum value of $f(x,y,z) = x^2 + y + z^2$ subject to
the condition that $g(x) = 2x + y + 4z = 6$.

****

$$\bar{\nabla} f(x, y, z) = \lambda \bar{\nabla} g(x, y, z)$$
$$(2x, 1, 2z) = \lambda(2, 1, 4) \Rightarrow \lambda = 1, \ x = 1, \ z = 2$$
$$\text{Then } g(x) = 6 \Rightarrow y = -4$$

$f(1, -4, 2) = 1$ is the required minimum.

Check: $F(x,z) = x^2 + 6 - 2x - 4z + z^2 = (x-1)^2 + (z-2)^2 + 1 \geq 1$.

3-106 ■■■

Find the maximum and minimum values of the function $f(x,y) = xy$
on the ellipse given by the equation

$$x^2 + \frac{y^2}{4} = 1$$

****

We use the method of Lagrange multipliers

$$\vec{\nabla} f = y \vec{i} + x \vec{j}$$

setting $g(x,y) = x^2 + \frac{y^2}{4} - 1 = 0$

$$\vec{\nabla} g = 2x \vec{i} + \tfrac{1}{2} y \vec{j}$$

Using the Lagrange multiplier λ and setting the components
of $\vec{\nabla} f$ and $\lambda \vec{\nabla} g$ equal, we have

Ⓐ $y = 2 \lambda x$ and $x = \tfrac{1}{2} \lambda y$ Ⓑ

or $\lambda = \frac{y}{2x}$ for $x \neq 0$ and $\lambda = \frac{2x}{y}$ for $y \neq 0$.

One solution of Ⓐ Ⓑ is $(0,0)$ which is not on the
ellipse.

substituting for λ in Ⓑ, Ⓐ we get:

$$\frac{y}{2x} = \frac{2x}{y} \quad \Rightarrow \quad y^2 = 4x^2$$

thus $g(x,y) = 0 \Rightarrow x^2 + \frac{(4x^2)}{4} - 1 = 0$

$$\Rightarrow \quad x = \pm \sqrt{\tfrac{1}{2}}$$

Since $y^2 = 4x^2$

$y = \pm 2x$

So we have four possible critical points:

$a = \left(\sqrt{\tfrac{1}{2}}, \tfrac{2}{\sqrt{2}} \right)$ $\qquad c = \left(-\sqrt{\tfrac{1}{2}}, \tfrac{2}{\sqrt{2}} \right)$

$b = \left(\sqrt{\tfrac{1}{2}}, -\tfrac{2}{\sqrt{2}} \right)$ $\qquad d = \left(-\sqrt{\tfrac{1}{2}}, -\tfrac{2}{\sqrt{2}} \right)$

$f(a) = 1$ $\qquad f(c) = -1$
$f(b) = -1$ $\qquad f(d) = 1$

The maximum value of $f(x,y)$ on the ellipse is 1, the minimum value is -1

━━━**3-107**

Find two positive numbers whose sum is eighteen and whose product is a maximum using the method of Lagrange Multipliers.

****

Call the numbers x and y. Then the objective equation is $f(x,y) = x \cdot y$ subject to the constraint equation $x+y=18$ or $x+y-18=0$. Also the Auxiliary equation $\mathcal{F}(x,y,\lambda) = f(x,y) + \lambda \cdot g(x,y)$ takes the form $\mathcal{F}(x,y,\lambda) = xy + \lambda(x+y-18)$, where λ is the Lagrange Multiplier. Now compute:

$\dfrac{\partial \mathcal{F}}{\partial x} = y + \lambda$

$\dfrac{\partial \mathcal{F}}{\partial y} = x + \lambda$ and solve

$\dfrac{\partial \mathcal{F}}{\partial \lambda} = x + y - 18$

① $y + \lambda = 0$
② $x + \lambda = 0$
③ $x + y - 18 = 0$

to find critical values. From ① and ② $\lambda = -y = -x$ or $x = y$. Sub. into ③ yields $2y - 18 = 0$ or $y_c = 9 = x_c$. Hence the numbers are $\underline{9}$ and $\underline{9}$.

3-108 ■■■■■■■■■■■■■■■■■■■■■■■■■■■■■■■■■■■■■■

Use the method of Lagrange multipliers to find points on the surface $x^2 + y^2 + z^2 = 3$ where the function $f(x,y,z) = x + y + z$ has:

(a) a minimum.

(b) a maximum.

****

LET $\quad g(x,y,z) = x^2 + y^2 + z^2 - 3.$

SOLVE THE SYSTEM: $\begin{cases} \nabla f + \lambda \nabla g = 0 \\ g(x,y,z) = 0. \end{cases}$

$\nabla f = \left(\dfrac{\partial f}{\partial x}, \dfrac{\partial f}{\partial y}, \dfrac{\partial f}{\partial y} \right) = (1, 1, 1)$

$\nabla g = \left(\dfrac{\partial g}{\partial x}, \dfrac{\partial g}{\partial y}, \dfrac{\partial g}{\partial z} \right) = (2x, 2y, 2z)$

$\therefore \; (1,1,1) + \lambda(2x, 2y, 2z) = 0$

$1 + 2\lambda x = 0 \implies x = -\dfrac{1}{2\lambda}$

$1 + 2\lambda y = 0 \implies y = -\dfrac{1}{2\lambda}$

$1 + 2\lambda z = 0 \implies z = -\dfrac{1}{2\lambda}$

BUT $\quad x^2 + y^2 + z^2 - 3 = 0.$

$\therefore \; \dfrac{1}{4\lambda^2} + \dfrac{1}{4\lambda^2} + \dfrac{1}{4\lambda^2} = 3$

$12\lambda^2 = 3$

$\lambda^2 = \dfrac{1}{4}$

$\therefore \; \lambda = \pm\dfrac{1}{2}$

$\therefore$ (a) A MINIMUM AT $(-1, -1, -1)$.

(b) A MAXIMUM AT $(1, 1, 1)$.

MISCELLANEOUS PROBLEMS

━━━━━━━━━━━━━━━━━━━━━━━━━━━━━━━━━━━━**3-109**

Given that f is differentiable at $P_0 = (x_0, y_0)$, it may be concluded
that (a) f is continuous at P_0 (b) the partial derivatives $D_1 f(P_0)$ and
$D_2 f(P_0)$ both exist (c) the directional derivative $D_{\vec{u}} f(P_0)$ exists for <u>all</u>
unit vectors $\vec{u}$ (d) the surface z = f(x,y) has a tangent plane at the
point $(x_0, y_0, f(x_0, y_0))$ (e) each of the preceding answers are correct.

****

*Each of (a) — (d) are consequences of differentiability,
so (e) is the best answer.*

━━━━━━━━━━━━━━━━━━━━━━━━━━━━━━━━━━━━**3-110**

The radius of a right circular cylinder is increasing at a rate of 2 cm/min
and the height is decreasing at 4 cm/min. At what rate is the volume
changing at the instant when the radius is 4cm and the height is 10 cm?

****

$$V = \pi r^2 h.$$

$$\frac{dV}{dt} = \pi \, 2r \, \frac{dr}{dt} h + \pi r^2 \frac{dh}{dt}$$

but $\frac{dr}{dt} = 2$; $\frac{dh}{dt} = -4$; $r = 4$, $h = 10$

so

$$\frac{dV}{dt} = \pi (8)(2)(10) + \pi (16)(-4)$$

$$= (160 - 64)\pi = 96\pi.$$

3-111 ━━━━━━━━━━━━━━━━━━━━━━━━━━━━━━━━━━━

Let f: $R^2 \to R$ be the function
$$f(x,y) = \begin{cases} 1-x^2 & \text{if } 0 \le x \le 1 \\ 0 & \text{otherwise} \end{cases}$$

a) Sketch the 2-dimensional sections {y = 0}, {x = 0}, {y = 1}, {x = 1}.
b) Sketch the graph of f. For clarity you can restrict the graph to the region $0 \le y \le 1$.

****

a) {y=0}: We wish to graph the fn of one variable
$$g(x) \equiv f(x,0) = \begin{cases} 1-x^2 & 0 \le x \le 1 \\ 0 & \text{otherwise} \end{cases}$$

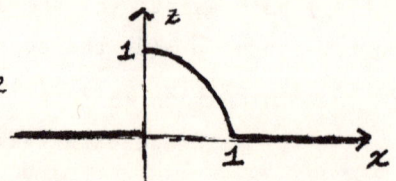

{y=1}: We wish to graph $g(x) \equiv f(x,1) = \begin{cases} 1-x^2 & 0 \le x \le 1 \\ 0 & \text{otherwise} \end{cases}$
Since this function is identical to the previous one, the section {y=1} looks exactly like the section {y=0} sketched above.

{x=0}: we wish to graph $g(y) \equiv f(0,y) = 1-0^2 = 1$
(For x=0 we use the top rule in the definition of f)

The graph is just

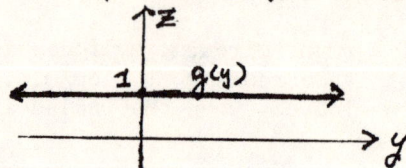

{x=1}: we wish to graph $g(y) \equiv f(1,y) = 1-1^2 = 0$

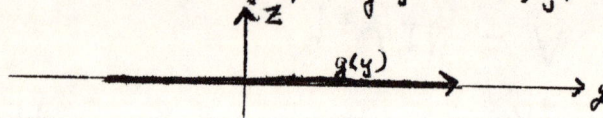

b) Using the information above plus more sections if necessary we see the graph looks like a sort of tunnel, with a vertical wall along the plane {x=0}

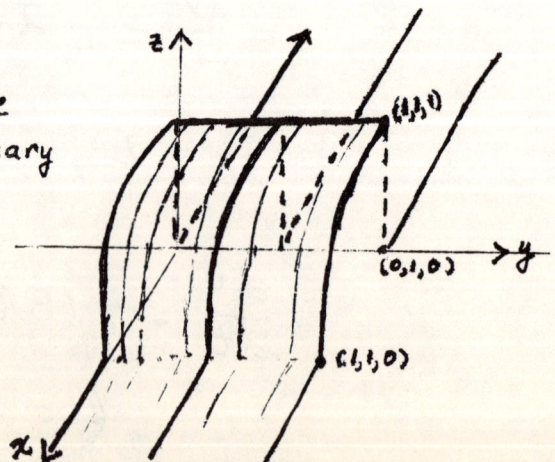

4

MULTIPLE INTEGRALS

INTERATED INTEGRALS IN THE PLANE

--- 4-1

Evaluate: $\int_1^2 \int_0^{\pi/x} x^2 \sin xy \, dy dx$

****

Begin with $\int_0^{\pi/x} x^2 \sin xy \, dy$. Treating x as a constant say

$x = k$ we find the anti-derivative as follows:

$\int k^2 \sin ky \, dy = k^2 \int \sin ky \, dy = k^2 \int \sin ky \frac{(k \, dy)}{k} = \frac{k^2}{k} \int \sin ky \, (k \, dy)$

or $\int k^2 \sin ky \, dy = k(-\cos ky) = -k \cos ky = -x \cos xy$. Hence

$\int_0^{\pi/x} x^2 \sin xy \, dy = -x \cos xy \Big|_0^{\pi/x} = -x \cos \left[x \left(\frac{\pi}{x} \right) \right] + x \cos 0$ or

$\int_0^{\pi/x} x^2 \sin xy \, dy = -x \cos \pi + x \cos 0 = -x(-1) + x(1) = 2x$. Now

$\int_1^2 \left(\int_0^{\pi/x} x^2 \sin xy \, dy \right) dx = \int_1^2 2x \, dx = x^2 \Big|_1^2 = 4 - 1 = \underline{3}$.

4-2 ██

Evaluate $\displaystyle\int_0^2 \int_{x^2}^4 (2x^2y + 7)\, dy\, dx$.

****

$$\int_{x^2}^4 (2x^2y + 7)\, dy = (x^2y^2 + 7y)\Big|_{y=x^2}^{y=4}$$

$$= (16x^2 + 28) - (x^6 + 7x^2) = 28 + 9x^2 - x^6$$

$$\int_0^2 (28 + 9x^2 - x^6)\, dx = (28x + 3x^3 - \frac{x^7}{7})\Big|_0^2$$

$$= 56 + 24 - \frac{128}{7} = \frac{432}{7}$$

4-3 ██

Assume f is continuous. Prove $\left[\int_a^b f(x)\, dx\right] \cdot \left[\int_c^d f(y)\, dy\right] = \iint_R f(x) \cdot f(y)\, dA$ where R is the rectangle bounded by $x = a$, $x = b$, $y = c$, $y = d$.

****

Let F be an antiderivative of f. Then $\left[\int_a^b f(x)\,dx\right]\left[\int_c^d f(y)\,dy\right]$ $= \left[F(b) - F(a)\right]\cdot\left[F(d) - F(c)\right]$. Also, $\iint_R f(x)\cdot f(y)\,dA =$ $\int_c^d\int_a^b f(x)f(y)\,dx\,dy = \int_c^d f(y)\cdot F(x)\Big|_a^b\,dy = \left[F(b)- F(a)\right]\cdot\int_c^d f(y)\,dy$ $= \left[F(b) - F(a)\right]\cdot\left[F(d) - F(c)\right]$.

■■**4-4**

Evaluate $\displaystyle\int_{0}^{\frac{\pi}{2}} \int_{0}^{\sin y} e^{x} \cos y \; dx \; dy$

****

$$\int_{0}^{\frac{\pi}{2}} \int_{0}^{\sin y} e^{x} \cos y \; dx \; dy = \int_{0}^{\pi/2} e^{x} \cos y \Big]_{x=0}^{x=\sin y} dy$$

$$= \int_{0}^{\pi/2} \left(e^{\sin y} \cos y - e^{0} \cos y \right) dy$$

$$= \int_{0}^{\pi/2} \left(e^{\sin y} \cos y - \cos y \right) dy = e^{\sin y} - \sin y \Big]_{0}^{\pi/2}$$

$$= \left(e^{\sin \pi/2} - \sin \frac{\pi}{2} \right) - \left(e^{\sin 0} - \sin 0 \right)$$

$$= (e - 1) - (1 - 0) = e - 2$$

4-5 ■■

$\int_0^1 \int_0^x F(x,y)\ dy\ dx$ is equivalent to (a) $\int_0^1 \int_y^1 F(x,y)\ dx\ dy$ (b) $\int_0^1 \int_1^y F(x,y)\ dx\ dy$
(c) $\int_0^1 \int_0^1 F(x,y)\ dx\ dy$ (d) $\int_0^1 \int_{-y}^y F(x,y)\ dx\ dy$ (e) $\int_0^1 \int_0^y F(x,y)\ dx\ dy$.

****

$\int_0^1 \int_0^x F(x,y)\ dy\ dx = \iint_R F(x,y)\ dA,$

where R is bounded below + above by

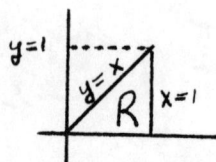

curves $y = 0$, $y = x$ and left + right by vertical lines

$x = 0$, $x = 1$, ie. R is the triangle shown above. To

interchange the order of integration, view R as bounded

left + right by curves $x = y$, $x = 1$ and below + above

by horizontal lines $y = 0$, $y = 1$. Hence also

$\iint_R F(x,y)\ dA = \int_0^1 \int_y^1 F(x,y)\ dx\ dy.$

THE DOUBLE INTEGRAL

■■■**4-6**

Express the integral $\int_0^2 \int_{x^2}^4 (xy^2 + x)\ dy\ dx$ as an

equivalent integral with the order of integration reversed.

****

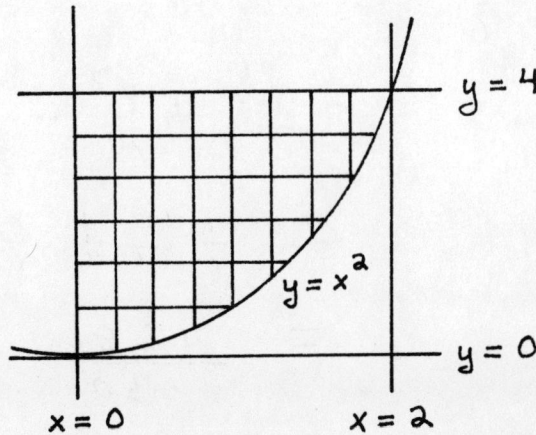

The region of
integration is shown
in the diagram.

Note: 1st quadrant, so

$$y = x^2 \longleftrightarrow x = \sqrt{y}$$

Answer: $\int_0^4 \int_0^{\sqrt{y}} (xy^2 + x)\ dx\ dy$

4-7

Evaluate $\int_0^1 \int_0^{x^3} (x^2 + y - y^2)\, dy\, dx$ and describe the region of integration.

****

$$\int_0^1 \int_0^{x^3} (x^2 + y - y^2)\, dy\, dx = \int_0^1 x^2 y + \frac{y^2}{2} - \frac{y^3}{3} \Big|_0^{x^3} dx$$

$$= \int_0^1 \left(x^5 + \frac{x^6}{2} - \frac{x^9}{3} \right) dx$$

$$= \frac{x^6}{6} + \frac{x^7}{14} - \frac{x^{10}}{30} \Big|_0^1$$

$$= \frac{1}{6} + \frac{1}{14} - \frac{1}{30}$$

$$= \frac{129}{630}$$

R is bounded below by $y=0$, above by $y=x^3$, on the left by $x=0$ and on the right by $x=1$. (See FIGURE)

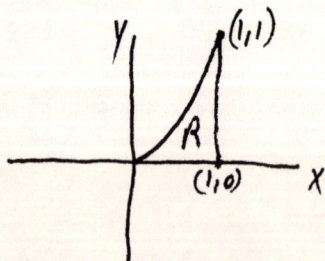

■■■ **4-8**

Change the order of integration in the following integral and evaluate:

$$\int_0^9 \int_{\sqrt{y}}^3 \sin(\pi x^3) \, dx \, dy$$

****

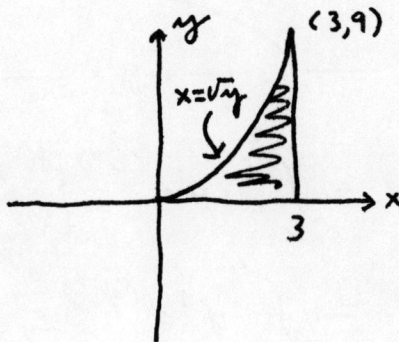

$$\int_0^9 \int_{\sqrt{y}}^3 \sin(\pi x^3) \, dx \, dy = \int_0^3 \int_0^{x^2} \sin(\pi x^3) \, dy \, dx$$

$$= \int_0^3 y \sin(\pi x^3) \Big|_0^{x^2} \, dx = \int_0^3 x^2 \sin(\pi x^3) \, dx$$

$$= -\frac{\cos(\pi x^3)}{3\pi} \Big|_0^3 = -\frac{1}{3\pi}\left[\cos 27\pi - \cos 0\right]$$

$$= -\frac{1}{3\pi}\left[-1-1\right] = \frac{2}{3\pi}$$

4-9 ■■

Evaluate the double integral of $ye^{(y^4)}$ over the region bounded by $y = \sqrt{x}$, $y = 2$, and $x = 0$.

****

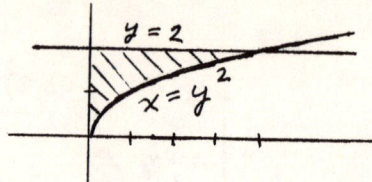

$$\int_0^2 \int_0^{y^2} ye^{(y^4)}\,dx\,dy = \int_0^2 \left(xye^{(y^4)} \Big|_{x=0}^{x=y^2} \right) dy$$

$$= \int_0^2 y^3 e^{(y^4)}\,dy$$

$$= \frac{1}{4} \int_0^2 e^{(y^4)} 4y^3\,dy$$

$$= \frac{1}{4} e^{(y^4)} \Big|_0^2$$

$$= \frac{1}{4} e^{16} - \frac{1}{4} e^0$$

$$= \frac{e^{16} - 1}{4}$$

■■ **4-10**

Evaluate the double integral

$$\int_0^1 \int_x^1 \sin y^2 \; dy \, dx$$

****

Note that we are unable to integrate $\int_x^1 \sin y^2 \, dx$, so we try reversing the order of integration.

Since the given order is $dy\,dx$, the lower and upper boundaries of the region are the graphs of $y=x$ and $y=1$ respectively, where $0 \le x \le 1$.

A sketch of the region of integration:

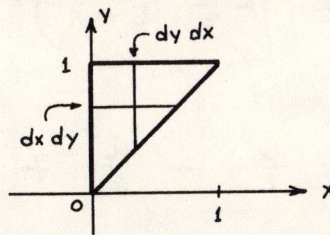

When the order is reversed, the left and right hand boundaries are given by $x=0$ and $x=y$ respectively, where $0 \le y \le 1$.

So we have

$$\int_0^1 \int_x^1 \sin y^2 \, dy \, dx \;=\; \int_0^1 \int_0^y \sin y^2 \, dx \, dy$$

$$=\; \int_0^1 y \sin y^2 \, dy$$

$$=\; \tfrac{1}{2}\left(-\cos y^2\right)\Big]_0^1$$

$$=\; \tfrac{1}{2}\left(1 - \cos 1\right)$$

4-11 ■■■

Evaluate the following double integral: $\int_0^a \int_0^{\sqrt{a^2-x^2}} (x+y)\,dy\,dx$.

****

$$\int_0^a \int_0^{\sqrt{a^2-x^2}} (x+y)\,dy\,dx = \int_0^a \left[xy + \frac{y^2}{2}\right]_0^{\sqrt{a^2-x^2}} dx$$

$$= \int_0^a \left[x(a^2-x^2)^{\frac{1}{2}} + \frac{a^2-x^2}{2}\right] dx$$

$$= \left[-\frac{1}{3}(a^2-x^2)^{\frac{3}{2}} + \frac{a^2 x}{2} - \frac{x^3}{6}\right]_0^a$$

$$= \frac{a^3}{2} - \frac{a^3}{6} + \frac{a^3}{3}$$

$$= \frac{3a^3 - a^3 + 2a^3}{6}$$

$$= \frac{2a^3}{3}$$

━━━━━━━━━━━━━━━━━━━━━━━━━━━━━━━━ 4-12

Use the change of variables   u = 2x - y,   v = x + y   to evaluate

$\iint_R$ (6x - 3y) dA   where   R   is the region bounded by   2x - y = 1,

2x - y = 3, x + y = 1,   x + y = 2.

****

The general change of variables formula for double integrals
(under suitable hypotheses) says that if $x = x(u,v)$, $y = y(u,v)$,
then $\iint_R f(x,y)\, dA = \iint_B f\big(x(u,v),\, y(u,v)\big) \cdot |J|\, dA$, where
B is the region in the uv plane corresponding to R,
and the Jacobian J is given by $J = \begin{vmatrix} \frac{\partial x}{\partial u} & \frac{\partial x}{\partial v} \\ \frac{\partial y}{\partial u} & \frac{\partial y}{\partial v} \end{vmatrix}$.

For this problem, the region B is the rectangle bounded
by $u = 1$ (from $2x - y = 1$), $u = 3$, $v = 1$, $v = 3$.

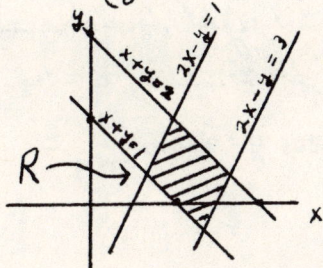

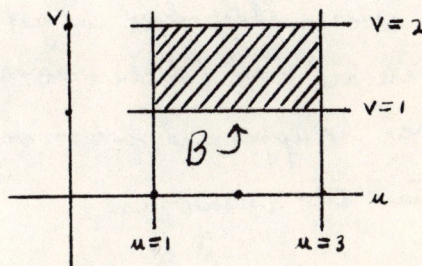

Also, in terms of u and v, the integrand is $6x - 3y = 3(2x-y) = 3u$.
To find J, we may use the fact that $\frac{1}{J} = \begin{vmatrix} \frac{\partial u}{\partial x} & \frac{\partial u}{\partial y} \\ \frac{\partial v}{\partial x} & \frac{\partial v}{\partial y} \end{vmatrix}$

$= \begin{vmatrix} 2 & -1 \\ 1 & 1 \end{vmatrix} = 3$, so $J = \frac{1}{3}$. Hence $\iint_R (6x-3y)\, dA =$

$\iint_B (3u) \cdot \frac{1}{3}\, dA = \int_1^3 \int_1^2 u\, dv\, du = \int_1^3 uv \Big|_1^2\, du = \int_1^3 u\, du = 4.$

4-13 ■■

Write $\displaystyle\int_0^{16}\int_0^{\sqrt{x}} f(x,y)\,dy\,dx$ with the order of integration reversed.

****

It is helpful to sketch the region of integration. The lower and upper boundaries are $y=0$ and $y=\sqrt{x}$, while the right and left boundaries are $x=16$ and $x=0$. We get

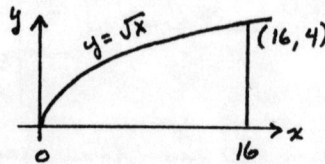

To reverse this, first we get the left and right boundaries. These are $x=y^2$ (because we have to solve for x) and $x=16$. Then the upper and lower boundaries are $y=4$ and $y=0$. Thus the answer is

$$\int_0^4 \int_{y^2}^{16} f(x,y)\,dx\,dy$$

4-14

The region D in R^2 shown below is bounded by $x = 1$, $y = e^x$, and $y = 1-x^2$.

a) Compute $\iint\limits_{D} x \, dA$ by finding $\int\limits_{x=0}^{1} \int\limits_{y=1-x^2}^{y=e^x} x \, dy \, dx$.

b) Write down the integral or integrals needed to compute $\iint\limits_{D} x \, dA$ with the order of integration reversed.

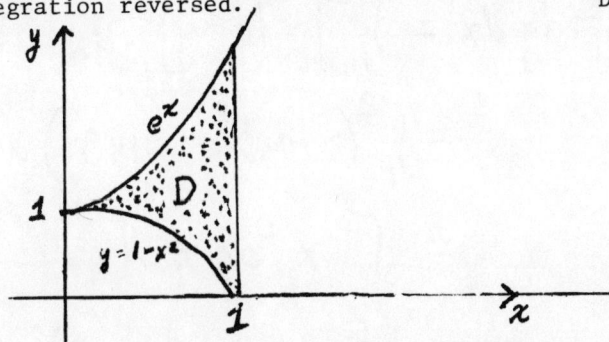

****

a) $\int_{0}^{1} \left[\int_{y=1-x^2}^{y=e^x} x \, dy \right] dx = \int_{x=0}^{1} x \left(y \big]_{1-x^2}^{e^x} \right) dx = \int_{0}^{1} xe^x - x + x^3 \, dx$

$\left(\begin{array}{c} \text{integration} \\ \text{by parts} \\ \text{on } 1^{0} \text{ integral !} \end{array} \right) \rightarrow = xe^x \big]_{0}^{1} - e^x \big]_{0}^{1} - \frac{x^2}{2} \big]_{0}^{1} + \frac{x^4}{4} \big]_{0}^{1}$

$= (e - 0) - (e - 1) - \frac{1}{2} + \frac{1}{4} = \boxed{\frac{3}{4}}$

b) This requires horizontal slices & we have to break D into 2 regions, above and below the line $y = 1$.

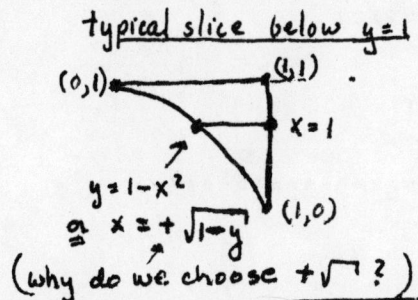

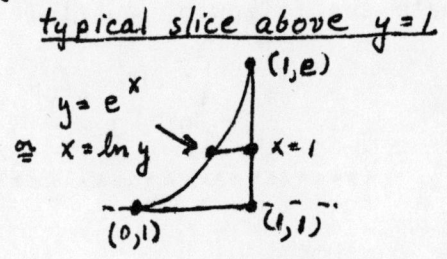

typical slice below $y=1$ typical slice above $y=1$

(why do we choose $+\sqrt{\ }$?)

So $\iint\limits_{D} x \, dA = \int\limits_{y=0}^{1} \int\limits_{x=\sqrt{1-y}}^{x=1} x \, dx \, dy + \int\limits_{y=1}^{e} \int\limits_{x=\ln y}^{x=1} x \, dx \, dy$

4-15 ■■

Evaluate $\int_1^2 \int_1^{x^2} \frac{x}{y} \, dy \, dx$

****

$$\int_1^2 \int_1^{x^2} \frac{x}{y} \, dy \, dx = \int_1^2 \left[x \log y \right]_1^{x^2} dx$$

$$= \int_1^2 \left(x \log x^2 - x \log 1 \right) dx$$

$$= \int_1^2 x \log x^2 \, dx$$

$$= \frac{1}{2} \left[x^2 \log x^2 - x^2 \right]_1^2$$

$$= \frac{1}{2} \left(4 \log 4 - 4 - 1 \cdot \log 1 + 1 \right)$$

$$= 2 \log 4 - \frac{3}{2}$$

4-16 ■■■

Evaluate the following multiple integral

$$\int_0^2 \int_0^y x^2 y^4 \, dx \, dy$$

****

$$\int_0^2 \int_0^y x^2 y^4 \, dx \, dy = \int_0^2 \frac{x^3 y^4}{3} \bigg|_0^y \, dy = \int_0^2 \frac{y^7}{3} \, dy$$

$$= \frac{y^8}{24} \bigg|_0^2 = \frac{256}{24} = \frac{32}{3}$$

■■■ **4-17**

Sketch a typical region bounded by the curves y = sin 2 x and y = cos x between consecutive points of intersection on the x-axis and use double integration to find the area of the region.

****

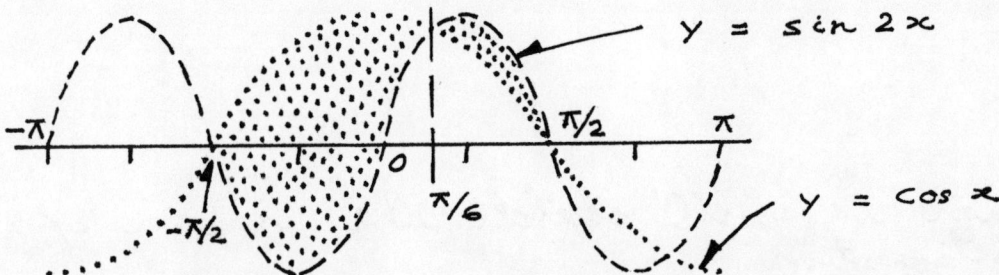

The curves $y = \sin 2x$ and $y = \cos x$ is shown above, where the typical region bounded by the two curves between consecutive points of intersection on the x-axis is shown shaded.

Using the method of double integration, we write:

$$\iint_D dA = \iint_{D_1} dA + \iint_{D_2} dA$$

$$= \int_{-\pi/2}^{\pi/6} \int_{\sin 2x}^{\cos x} dy\, dx + \int_{\pi/6}^{\pi/2} \int_{\cos x}^{\sin 2x} dy\, dx$$

$$= \int_{-\pi/2}^{\pi/6} (\cos x - \sin 2x)\, dx + \int_{\pi/6}^{\pi/2} (\sin 2x - \cos x)\, dx$$

$$= \sin x \Big|_{-\pi/2}^{\pi/6} - \left(-\frac{\cos 2x}{2}\right)\Big|_{-\pi/2}^{\pi/6}$$

$$+ \left(-\frac{\cos 2x}{2}\right)\Big|_{\pi/6}^{\pi/2} - (\sin x)\Big|_{\pi/6}^{\pi/2}$$

$$= \left[\frac{1}{2} - (-1)\right] + \left[\frac{\frac{1}{2} - (-1)}{2}\right] - \left[\frac{-1 - \frac{1}{2}}{2}\right] - \left[1 - \frac{1}{2}\right]$$

$$= \frac{1}{2} + 1 + \frac{3/2}{2} + \frac{3/2}{2} - \frac{1}{2} = \frac{5}{2} \quad \underline{Ans.}$$

4-18 ▬▬▬▬▬▬▬▬▬▬▬▬▬▬▬▬▬▬▬▬▬▬▬▬

Find the missing bounds of $\int_0^9 \int_{\sqrt{y}}^3 f(x,y)dxdy = \int\int f(x,y)dydx$.

****

First we will sketch the region of integration.

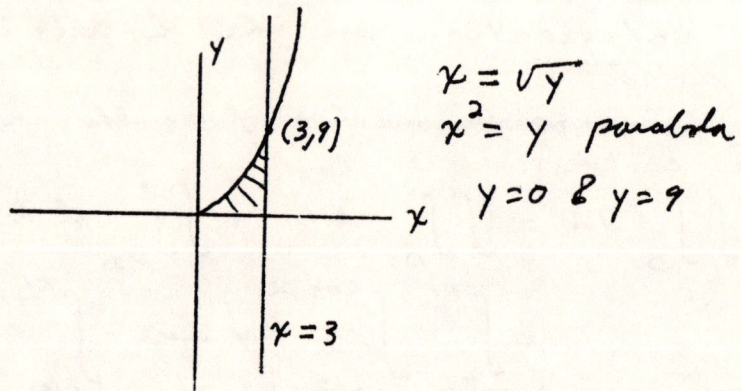

$x = \sqrt{y}$
$x^2 = y$ parabola
$y = 0 \ \& \ y = 9$

$$\int_0^9 \int_{\sqrt{y}}^3 f(x,y)dxdy = \int_0^3 \int_0^{x^2} f(x,y)dy\,dx.$$

AREA BY DOUBLE INTEGRALS

■■■**4-19**

Write a double integral which measures the area of the ellipse

$$\frac{x^2}{4} + \frac{y^2}{9} = 1.$$

****

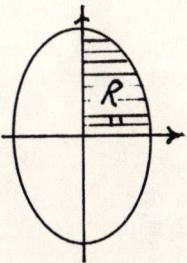

$$x = \pm\sqrt{4 - \frac{4y^2}{9}} = \pm\frac{2}{3}\sqrt{9 - y^2}$$

$$A = 4\int_0^3\int_0^{\frac{2}{3}\sqrt{9-y^2}} dx\,dy = \iint_R 4\,dA$$

If an evaluation of the above double integral were required, A could be found as follows.

$$A = 4\int_0^3 x\,\Big|_0^{\frac{2}{3}\sqrt{9-y^2}} dy = 4\int_0^3 \frac{2}{3}\sqrt{9-y^2}\,dy. \quad \text{Let } y = 3\sin u$$

$$A = \frac{8}{3}\int_0^{\pi/2}\sqrt{9 - 9\sin^2 u}\,\,3\cos u\,du = 24\int_0^{\pi/2}\cos^2 u\,du$$

$$= 12\int_0^{\pi/2}(1 + \cos 2u)\,du = 12\Big[u + \frac{1}{2}\sin 2u\Big]_0^{\pi/2} = 6\pi$$

As a check, $A = \pi r_1 r_2 = \pi \cdot 2 \cdot 3 = 6\pi.$

4-20 ■■■

Find the area between $y = 2x^2 - 4$ and $y = x^2 - 2x - 1$.

****

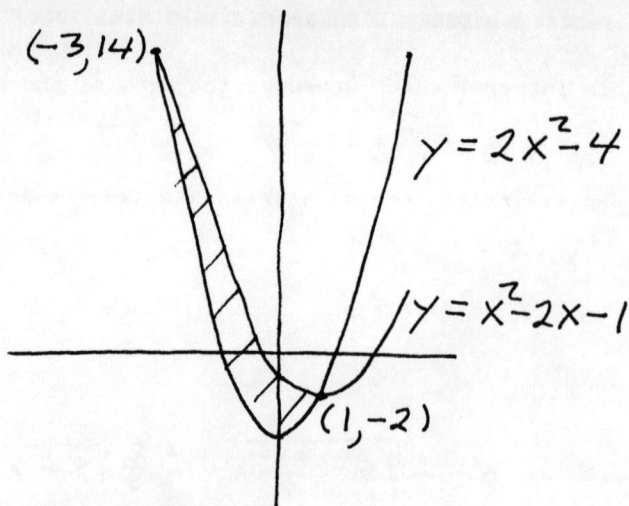

Solving simultaneously, $2x^2 - 4 = x^2 - 2x - 1$,

$x^2 + 2x - 3 = 0$, $(x+3)(x-1) = 0$, $x = 1$ or -3.

$$\text{Area} = \int_{-3}^{1} \int_{2x^2-4}^{x^2-2x-1} dy\, dx$$

$$= \int_{-3}^{1} x^2 - 2x - 1 - (2x^2 - 4)\, dx$$

$$= \int_{-3}^{1} -x^2 - 2x + 3\, dx \quad = -\frac{x^3}{3} - x^2 + 3x \Big|_{-3}^{1}$$

$$= -\frac{1}{3} - 1 + 3 - 9 + 9 + 9$$

$$= 10\frac{2}{3}.$$

━━━━━━━━━━━━━━━━━━━━━━━━━━━━━━━━━━━━━━ **4-21**

The area of a certain region R in the xy-plane is expressed by the iterated integral $\int_0^2 \int_{x^2}^{x+2} dy\, dx$. (a) Draw a careful sketch of the region R, and (b) express the area of R reversing the order of integration.

****

a)

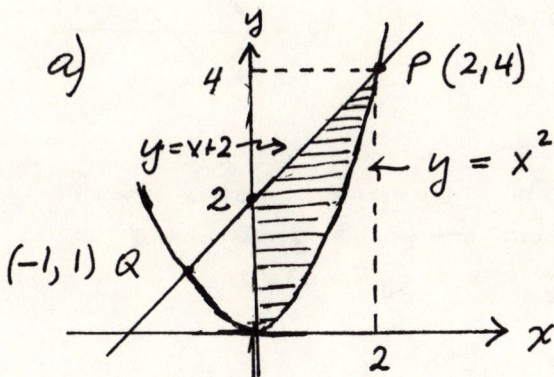

R is the hatched region

the coordinates of P and Q are found from: $x^2 = x + 2$

$x^2 - x - 2 = 0$

$(x - 2)(x + 1) = 0$

$x = 2 \ \& \ x = -1$

b) to reverse the order of integration, we need to solve for x in terms of y:

$y = x^2 \longrightarrow x = \sqrt{y}$ (we need the positive branch of the parabola only)

$y = x + 2 \longrightarrow x = y - 2$

From the picture in (a) it is now clear that the appropriate iterated integrals are:

$$\int_0^2 \int_0^{\sqrt{y}} dx\, dy + \int_2^4 \int_{y-2}^{\sqrt{y}} dx\, dy$$

4-22 ■■■

Using a double integral, find the area enclosed by the parabola $y^2 = 2x + 4$ and the line $y = x - 2$.

****

The graphs are shown at right. Find the points of intersection $(0, -2)$ and $(6, 4)$ by solving the equations simultaneously. It will be easier to integrate on x first and y second. For this purpose, notice the equation of the line can be written $x = y + 2$, and the parabola can be written $x = (y^2 - 4)/2$.

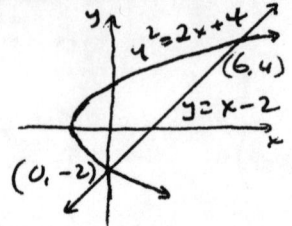

$$\text{Area} = \int_{-2}^{4} \int_{\frac{y^2-4}{2}}^{y+2} dx\, dy = \int_{-2}^{4} x \Big|_{\frac{y^2-4}{2}}^{y+2} dy$$

$$= \int_{-2}^{4} \left[(y+2) - \left(\frac{y^2-4}{2}\right)\right] dy = \int_{-2}^{4} \left(-\tfrac{1}{2}y^2 + y + 4\right) dy$$

$$= -\tfrac{1}{6}y^3 + \tfrac{1}{2}y^2 + 4y \Big|_{-2}^{4}$$

$$= -\tfrac{1}{6}(4)^3 + \tfrac{1}{2}(4)^2 + 4(4) - \left[-\tfrac{1}{6}(-2)^3 + \tfrac{1}{2}(-2)^2 + 4(-2)\right] = 18$$

If the order of integration is reversed, two integrals are needed, since the curves cross horizontally at $x = 0$.

$$\int_{-2}^{0} \int_{-\sqrt{2x+4}}^{\sqrt{2x+4}} dy\, dx + \int_{0}^{6} \int_{x-2}^{\sqrt{2x+4}} dy\, dx$$

When these integrals are evaluated, their sum is 18.

■■**4-23**

Use double integrals to find the area of the region in the first quadrant bounded by $y^2 = x^3$ and $y = x$.

****

Choose the order of integration to be $dx\, dy$; hence, we must draw the representative rectangle parallel to the x-axis. See the diagram below.

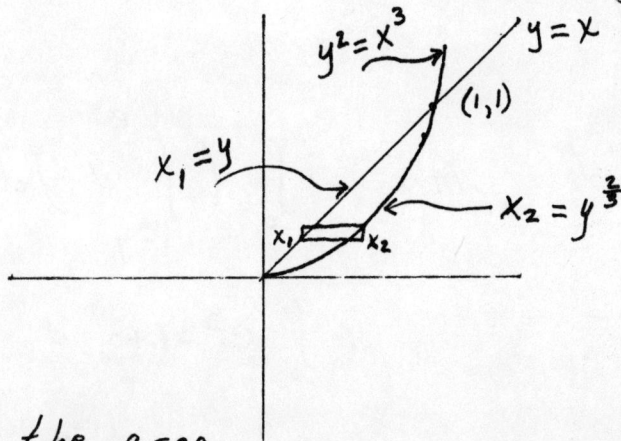

Let A be the area.

$$A = \int_0^1 \int_{x_1 = y}^{x_2 = y^{\frac{2}{3}}} dx\, dy = \int_0^1 x \Big|_{x_1 = y}^{x_2 = y^{\frac{2}{3}}} dy$$

$$= \int_0^1 (y^{\frac{2}{3}} - y)\, dy$$

$$= \left(\frac{3}{5} y^{\frac{5}{3}} - \frac{y^2}{2}\right)\Big|_0^1$$

$$= \frac{1}{10}$$

4-24

Find the area of the region bounded by the curves $y = \ln x$, $y = 1-x$ and $y = 1$.

****

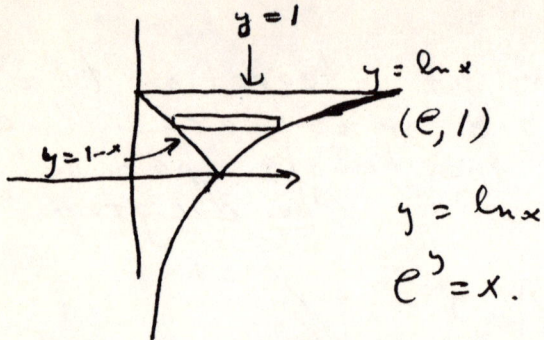

$y = \ln x$

$(e, 1)$

$y = \ln x$

$e^y = x.$

$y = 1-x.$

$x = 1-y$

$$Area = \int_0^1 \left(\int_{1-y}^{e^y} dx \right) dy$$

$$= \int_0^1 e^y - 1 + y \; dy$$

$$= e^y - y + \frac{y^2}{2} \Big|_0^1$$

$$= (e - 1 + \tfrac{1}{2}) - (1 - 0 + 0)$$

$$= e - \tfrac{3}{2}$$

4-25

Write $\int_{-5}^{0} \int_{-\sqrt{4-y}}^{y+2} dxdy + \int_{0}^{4} \int_{-\sqrt{4-y}}^{\sqrt{4-y}} dxdy$ as a single integral.

****

First we must sketch the regions of integration. I will indicate the region represented by the left double integral with ///// and the second with ⋰ .

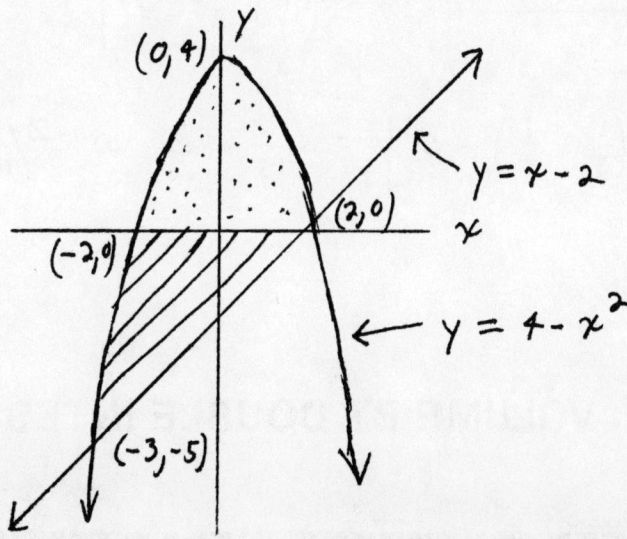

From $\int_{-5}^{0} \int_{-\sqrt{4-y}}^{y+2} dx\,dy$ ⟹

$x = y+2$ ⟹ $y = x-2$
$x = -\sqrt{4-y}$ ⟹ $y = 4-x^2$

and $\int_{0}^{4} \int_{-\sqrt{4-y}}^{\sqrt{4-y}} dx\,dy$ ⟹

$x = \sqrt{4-y}$ ⟹ $y = 4-x^2$
$x = -\sqrt{4-y}$ ⟹ $y = 4-x^2$

therefore as a single integral we have $\int_{-3}^{2} (4-x^2) - (x-2)\, dx$.

4-26 ▬▬▬▬▬▬▬▬▬▬▬▬▬▬▬▬▬▬▬▬▬▬▬▬▬▬

Find the area in the first quadrant between $y=x^2$ and $y=x^4$ using double integrals.

****

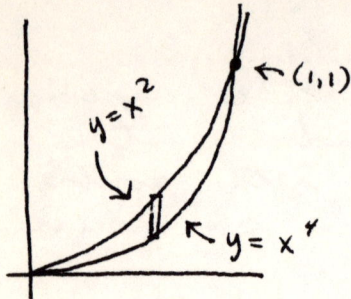

$$A = \int_0^1 \int_{x^4}^{x^2} 1 \, dy \, dx = \int_0^1 \left[y \right]_{x^4}^{x^2} dx$$

$$= \int_0^1 \left[x^2 - x^4 \right] dx = \left[\frac{x^3}{3} - \frac{x^5}{5} \right]_0^1$$

$$= \left[\left(\frac{1}{3} - \frac{1}{5} \right) - 0 \right] = \frac{5}{15} - \frac{3}{15} = \frac{2}{15}$$

VOLUME BY DOUBLE INTEGRALS

4-27 ▬▬▬▬▬▬▬▬▬▬▬▬▬▬▬▬▬▬▬▬▬▬▬▬▬▬

Find the volume under the surface $f(x,y) = xy$ over the region bounded by $y=2x$, $x=2y$, $x=0$ and $x=1$.

****

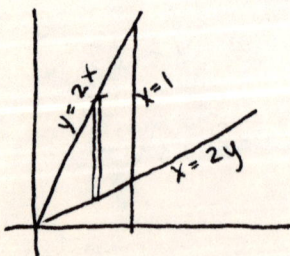

$$V = \int_0^1 \int_{x/2}^{2x} xy \, dy \, dx = \int_0^1 x \left[\frac{y^2}{2} \right]_{x/2}^{2x} dx$$

$$= \int_0^1 x \left[\frac{4x^2}{2} - \frac{x^2/4}{2} \right] dx = \int_0^1 \frac{15}{8} x^3 \, dx$$

$$= \frac{15}{8} \cdot \frac{x^4}{4} \Big|_0^1 = \frac{15}{8 \cdot 4} - 0 = \frac{15}{32}$$

■■■ **4-28**

Use a double integral to find the volume of the solid bounded above by $z = 9-x^2$, below by $z=0$, and laterally by $y^2 = 3x$.

****

$z = 9-x^2$ intersects the xy-plane along the lines $x = 3$ and $x = -3$.

∴ the base of the solid is as shown in the diagram.

Note: $y^2 = 3x \longleftrightarrow x = \dfrac{y^2}{3}$

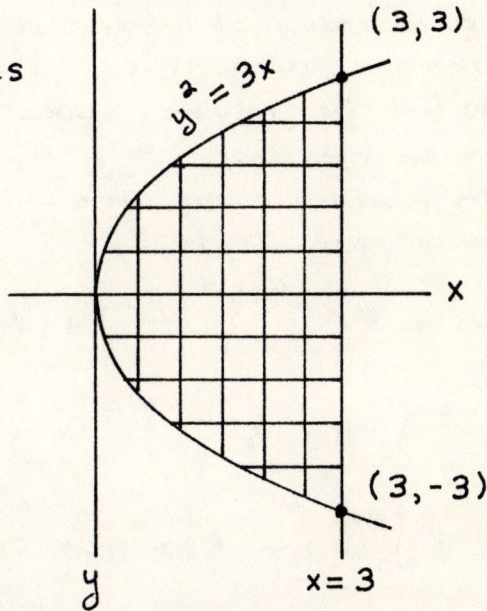

$$Vol = \int_{-3}^{3} \int_{\frac{y^2}{3}}^{3} (9-x^2)\, dx\, dy$$

$$\int_{\frac{y^2}{3}}^{3} (9-x^2)\, dx = \left(9x - \frac{x^3}{3} \right) \Bigg|_{\frac{y^2}{3}}^{3}$$

$$= (27-9) - \left(3y^2 - \frac{y^6}{81} \right) = 18 - 3y^2 + \frac{y^6}{81}$$

$$\int_{-3}^{3} \left(18 - 3y^2 + \frac{y^6}{81} \right) dy = \left(18y - y^3 + \frac{y^7}{567} \right) \Bigg|_{-3}^{3}$$

$$= \left(54 - 27 + \frac{27}{7} \right) - \left(-54 + 27 - \frac{27}{7} \right) = \frac{432}{7}$$

4-29 ■■

Find the volume of the solid in the first octant that is bounded by the plane $y + z = 4$, the cylinder $y = x^2$, and the xy and yz planes.

****

Several orders of integration are possible. We will use $\iint z\,dy\,dx$ to find the volume. Limits on y are from $y = x^2$ to $y = 4$, and the resulting slabs are summed on x from 0 to 2.

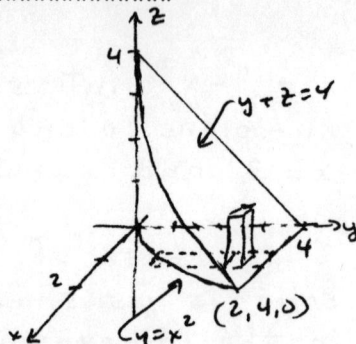

$$\text{Volume} = \int_0^2 \int_{x^2}^4 (4-y)\,dy\,dx$$

$$= \int_0^2 4y - \tfrac{1}{2}y^2 \Big|_{x^2}^4 dx = \int_0^2 \left[16 - \tfrac{1}{2}(16) - \left(4x^2 - \tfrac{1}{2}x^4\right)\right] dx$$

$$= \int_0^2 \left(8 - 4x^2 + \tfrac{1}{2}x^4\right) dx = 8x - \tfrac{4}{3}x^3 + \tfrac{1}{10}x^5 \Big|_0^2$$

$$= 8(2) - \tfrac{4}{3}(2)^3 + \tfrac{1}{10}(2)^5 = \frac{128}{15}.$$

If $\iint z\,dx\,dy$ is used, then volume $= \int_0^4 \int_0^{\sqrt{y}} (4-y)\,dx\,dy$.

If $\iint x\,dz\,dy$ is used, then volume $= \int_0^4 \int_0^{4-y} \sqrt{y}\,dz\,dy$.

Triple integrals can be used as well. In the order worked out in detail above,

$$\text{volume} = \int_0^2 \int_{x^2}^4 \int_0^{4-y} dz\,dy\,dx.$$

■■■■■■■■■■■■■■■■■■■■■■■■■■■■■■■■■■■■■■ **4-30**

Find the volume of the solid bounded by $z = 4-x^2$, $y^2 = 4x$ and the xy plane.

****

The region in the xy plane over which we integrate is the region in the xy-plane bounded by the 2 parabolic cylinders $z = 4-x^2$ and $y^2 = 4x$. Hence the region over which we integrate is the following:

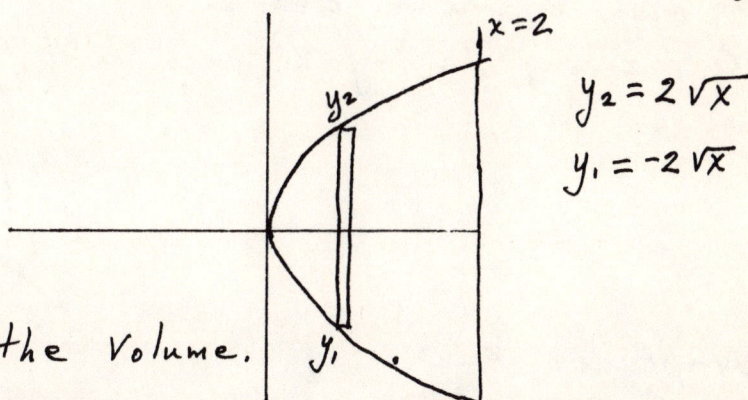

Let V be the volume.

Then

$$V = \int_0^2 \int_{y_1 = -2\sqrt{x}}^{y_2 = 2\sqrt{x}} (4-x^2)\, dy\, dx = 2\int_0^2 \int_0^{2\sqrt{x}} (4-x^2)\, dy\, dx$$

$$= 2\int_0^2 (4y - x^2 y)\Big|_0^{2\sqrt{x}} dx = 2\int_0^2 \left(8\sqrt{x} - 2x^{\frac{5}{2}}\right) dx$$

$$= 2\left[8\left(\tfrac{2}{3}\right)x^{\frac{3}{2}} - 2\left(\tfrac{2}{7}\right)x^{\frac{7}{2}}\right]_0^2$$

$$= \left(\tfrac{64}{3} - \tfrac{64}{7}\right)\left(2^{\frac{1}{2}}\right)$$

$$= \frac{256}{21}\sqrt{2}$$

4-31 ■■■■■■■■■■■■■■■■■■■■■■■■■■■■■■■■■■■■■

Let S be the surface defined by $z = 1-y-x^2$. Let V be the volume of the 3-dimensional region in the first octant bounded by S and the coordinate planes. Set up (but do <u>not</u> evaluate) the iterated integrals for V: (a) integrate first with respect to x and then y; (b) integrate first with respect to y and then x.

****

to draw a sketch, look at the traces of $z = 1-y-x^2$

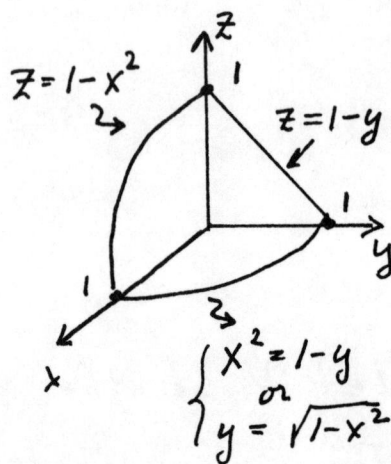

$$z = 1-x^2$$

$$z = 1-y$$

$$\begin{cases} x^2 = 1-y \\ \quad or \\ y = \sqrt{1-x^2} \end{cases}$$

xy-plane: $z = 0$

$$0 = 1-y-x^2 \quad or \quad x^2 = 1-y$$

(parabola)

yz-plane: $x = 0 \qquad z = 1-y$

(straight line)

xz-plane: $y = 0 \qquad z = 1-x^2$ (parabola)

(a) fix y first: $\displaystyle\int_0^1 \int_0^{\sqrt{1-y}} (1-y-x^2)\, dx\, dy$

(b) fix x first $\displaystyle\int_0^1 \int_0^{1-x^2} (1-y-x^2)\, dy\, dx$

■■ **4-32**

Find the volume of the solid under the surface $z = x^2 + y^2$ and lying above the region $\{(x,y) \mid 0 \le x \le 1, \ x^2 \le y \le \sqrt{x}\}$

****

$$VOLUME = \int_0^1 \int_{x^2}^{\sqrt{x}} (x^2 + y^2) \, dy \, dx$$

$$= \int_0^1 \left[\left(x^2 y + \frac{y^3}{3}\right)\right]_{x^2}^{\sqrt{x}} dx$$

$$= \int_0^1 \left(x^{5/2} + \frac{1}{3} x^{3/2} - x^4 - \frac{1}{3} x^6\right) dx$$

$$= \left[\frac{2}{7} x^{7/2} + \frac{2}{15} x^{5/2} - \frac{1}{5} x^5 - \frac{1}{21} x^7\right]_0^1$$

$$= \frac{2}{7} + \frac{2}{15} - \frac{1}{5} - \frac{1}{21} = \frac{6}{35}$$

DOUBLE INTEGRALS IN POLAR COORDINATES

■■■ **4-33**

If R is the region inside the circle $x^2 + y^2 = 4$, then $\iint_R x\sqrt{x^2 + y^2} \, dA$ is equal to (a) $\int_0^2 \int_0^{2\pi} r^2 \, d\theta \, dr$ (b) $\int_0^2 \int_0^{\pi/2} r^2 \cos\theta \, d\theta \, dr$ (c) $4\int_0^2 \int_0^{2\pi} r^3 \sin\theta \, d\theta dr$ (d) $\int_0^2 \int_0^{2\pi} r^3 \cos\theta \, d\theta \, dr$ (e) $\int_{-2}^2 \int_{-2}^2 x\sqrt{x^2 + y^2} \, dy \, dx$.

****

In polar coordinates, $x\sqrt{x^2+y^2} = (r\cos\theta) \cdot r = r^2 \cos\theta$.

Inserting the Jacobian for polar coordinates, r,

$$\iint_R x\sqrt{x^2+y^2} \, dA = \int_0^2 \int_0^{2\pi} r^3 \cos\theta \, d\theta \, dr.$$

4-34 ▪▪▪

Find the area inside the circle r = 2 cos θ and outside the circle r = 1.

****

$$2 \cos \theta = 1$$
$$\cos \theta = \frac{1}{2}$$
$$\theta = \pm \frac{\pi}{3}$$

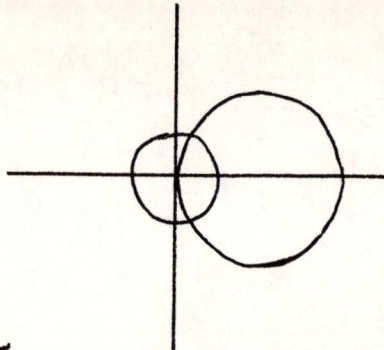

$$AREA = \int_{-\frac{\pi}{3}}^{\frac{\pi}{3}} \int_{1}^{2 \cos \theta} r \, dr \, d\theta$$

$$= \int_{-\frac{\pi}{3}}^{\frac{\pi}{3}} \left[\frac{r^2}{2} \right]_{1}^{2 \cos \theta} d\theta$$

$$= \int_{-\frac{\pi}{3}}^{\frac{\pi}{3}} \left(2 \cos^2 \theta - \frac{1}{2} \right) d\theta$$

$$= \int_{-\frac{\pi}{3}}^{\frac{\pi}{3}} \left(1 + \cos 2\theta - \frac{1}{2} \right) d\theta$$

$$= \left[\frac{1}{2} \theta + \frac{1}{2} \sin 2\theta \right]_{-\frac{\pi}{3}}^{\frac{\pi}{3}}$$

$$= \frac{\pi}{6} + \frac{1}{2} \sin \frac{2\pi}{3} + \frac{\pi}{6} - \frac{1}{2} \sin \left(-\frac{2\pi}{3} \right)$$

$$= \frac{\pi}{3} + \frac{\sqrt{3}}{2}$$

━━━━━━━━━━━━━━━━━━━━━━━━━━━━━━━━━━━━━━ **4-35**

Use polar coordinates to evaluate $\displaystyle\int_0^1 \int_0^{\sqrt{1-x^2}} e^{-(x^2+y^2)}\, dy\, dx.$

****

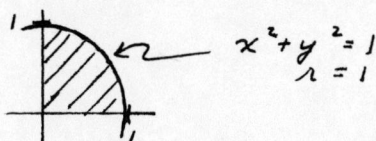

$x^2+y^2=1$
$r=1$

$$\int_0^{\frac{\pi}{2}} \int_0^1 e^{-r^2} r\, dr\, d\theta = \int_0^{\frac{\pi}{2}} \left(-\tfrac{1}{2}\right)\int_0^1 e^{-r^2}(-2r)\, dr\, d\theta$$

$$= \int_0^{\frac{\pi}{2}} \left(-\tfrac{1}{2} e^{-r^2}\Big|_{r=0}^{r=1}\right) d\theta$$

$$= \int_0^{\frac{\pi}{2}} \left(-\tfrac{1}{2} e^{-1} + \tfrac{1}{2} e^0\right) d\theta$$

$$= \int_0^{\frac{\pi}{2}} \frac{e-1}{2e}\, d\theta$$

$$= \frac{e-1}{2e}\, \theta \Big|_0^{\frac{\pi}{2}}$$

$$= \frac{e-1}{4e}\, \pi$$

4-36

Convert the following integral to polar coordinates and evaluate:

$$\int_0^2 \int_{-\sqrt{4-x^2}}^0 \frac{xy}{\sqrt{x^2 + y^2}} \, dy \, dx$$

****

$$\int_0^2 \int_{-\sqrt{4-x^2}}^0 \frac{xy}{\sqrt{x^2+y^2}} \, dy \, dx$$

$$= \int_{-\pi/2}^0 \int_0^2 \frac{r^2 \cos\theta \sin\theta}{r} \, r \, dr \, d\theta$$

$$= \int_{-\pi/2}^0 \frac{\cos\theta \sin\theta \, r^3}{3} \Big|_0^2 \, d\theta = \frac{8}{3} \int_{-\pi/2}^0 \cos\theta \sin\theta \, d\theta$$

$$= \frac{4}{3} \sin^2\theta \Big|_{-\pi/2}^0 = \frac{4}{3}\left[0 - 1\right] = -\frac{4}{3}$$

4-37

Use a double integral to find the area of the polar region
which lies inside the curve $r = 2 + 2 \sin \theta$ and outside the
curve $r = 3$.

****

$2 + 2 \sin \theta = 3 \implies \sin \theta = \frac{1}{2} \implies \theta = \frac{\pi}{6}, \frac{5\pi}{6}$

$$\therefore \quad \text{Area} = \int_{\frac{\pi}{6}}^{\frac{5\pi}{6}} \int_{3}^{(2+2\sin\theta)} r \, dr \, d\theta$$

$$\int_{3}^{2+2\sin\theta} r \, dr = \frac{1}{2} r^2 \Big|_{3}^{2+2\sin\theta}$$

$$= \frac{1}{2} \left(4 + 8 \sin \theta + 4 \sin^2 \theta - 9 \right) = 2 \sin^2 \theta + 4 \sin \theta - \frac{5}{2}$$

$$\int_{\frac{\pi}{6}}^{\frac{5\pi}{6}} \left(2 \sin^2 \theta + 4 \sin \theta - \frac{5}{2} \right) d\theta$$

$$= \int_{\frac{\pi}{6}}^{\frac{5\pi}{6}} \left((1 - \cos 2\theta) + 4 \sin \theta - \frac{5}{2} \right) d\theta$$

$$= \int_{\frac{\pi}{6}}^{\frac{5\pi}{6}} \left(- \cos 2\theta + 4 \sin \theta - \frac{3}{2} \right) d\theta$$

$$= \left(-\frac{1}{2} \sin 2\theta - 4 \cos \theta - \frac{3}{2} \theta \right) \Big|_{\frac{\pi}{6}}^{\frac{5\pi}{6}}$$

$$= \left(\frac{\sqrt{3}}{4} + 2\sqrt{3} - \frac{5\pi}{4} \right) - \left(-\frac{\sqrt{3}}{4} - 2\sqrt{3} - \frac{\pi}{4} \right)$$

$$= \frac{9\sqrt{3}}{2} - \pi$$

4-38

Show that $\int_{-\infty}^{\infty} e^{-x^2} dx = \sqrt{\pi}$.

****

Since e^{-x^2} is even we may conclude that $\int_0^\infty e^{-x^2} dx = \frac{\sqrt{\pi}}{2}$. Let $I = \int_0^\infty e^{-x^2} dx =$

$\int_0^\infty e^{-y^2} dy$. Now $I^2 = I \cdot I = \left(\int_0^\infty e^{-x^2} dx \right)\left(\int_0^\infty e^{-y^2} dy \right) =$

$\int_0^\infty \int_0^\infty e^{-(x^2+y^2)} dx dy = \int_0^{\frac{\pi}{2}} \int_0^\infty e^{-r^2} r \, dr \, d\theta =$

$-\frac{1}{2} \int_0^{\frac{\pi}{2}} \left(\lim_{t \to \infty} \int_0^t e^{-r^2} (-2r \, dr) \right) d\theta = -\frac{1}{2} \int_0^{\frac{\pi}{2}} \lim_{t \to \infty} e^{-r^2} \Big|_0^t d\theta =$

$-\frac{1}{2} \int_0^{\frac{\pi}{2}} \left(\lim_{t \to \infty} \left(\frac{1}{e^{t^2}} - 1 \right) \right) d\theta = +\frac{1}{2} \int_0^{\frac{\pi}{2}} d\theta = \frac{1}{2} \left\{ \theta \Big|_0^{\frac{\pi}{2}} \right\} =$

$\frac{\pi}{4}$. Clearly $I^2 = \frac{\pi}{4}$ and $I > 0 \Rightarrow I = \frac{\sqrt{\pi}}{2}$.

$\therefore \int_{-\infty}^\infty e^{-x^2} dx = \sqrt{\pi}$

■■■ **4-39**

Evaluate $\int_{-1}^{0} \int_{-\sqrt{1-y^2}}^{0} \cos(x^2 + y^2) \, dx \, dy$.

****

First we will sketch the region of integration.

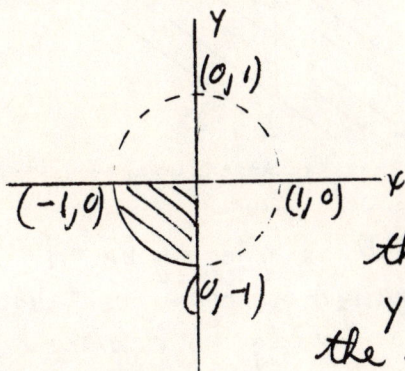

$x = -\sqrt{1-y^2}$

$x^2 = 1 - y^2$

$x^2 + y^2 = 1$ a circle

$x = -\sqrt{1-y^2} \Rightarrow$ the lower branch.

$y = 0 \ \& \ y = -1 \Rightarrow$ the lower left.

The $x^2 + y^2 \Rightarrow$ that we should change into Polar Coordinates.

$$\int_{-1}^{0} \int_{-\sqrt{1-y^2}}^{0} \cos(x^2+y^2) \, dx \, dy = \int_{\pi}^{\frac{3\pi}{2}} \int_{0}^{1} \cos r^2 (r \, dr \, d\theta) =$$

$$\int_{\pi}^{\frac{3\pi}{2}} \frac{1}{2} \int_{0}^{1} \cos r^2 (2r \, dr) \, d\theta = \int_{\pi}^{\frac{3\pi}{2}} \frac{1}{2} \sin r^2 \Big|_{0}^{1} \, d\theta =$$

$$\frac{1}{2} \int_{\pi}^{\frac{3\pi}{2}} (\sin 1 - \sin 0) \, d\theta = \frac{\sin 1}{2} \int_{\pi}^{\frac{3\pi}{2}} d\theta = \frac{\sin 1}{2} \left\{ \theta \Big|_{\pi}^{\frac{3\pi}{2}} \right\} =$$

$$\frac{\sin 1}{2} \left\{ \frac{3\pi}{2} - \pi \right\} = \frac{(\sin 1)\pi}{4} .$$

4-40 ∎∎

Let D be the exterior of the unit disc in R^2. (See picture below).

Find $\iint\limits_{D} e^{-(x^2 + y^2)}$ dx dy. Hint: use polar coordinates.

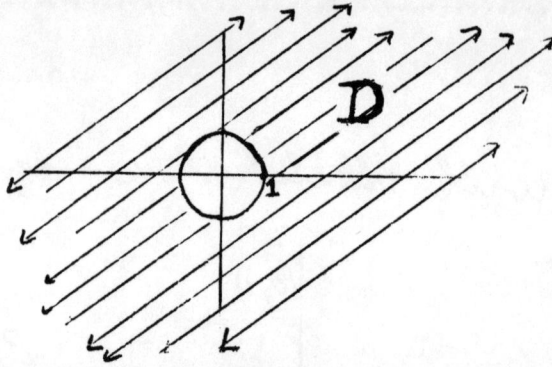

****

In polar coordinates D is the region $\{r \geq 1\}$.
The Jacobean for changing from rectangular to polar coordinates is r, so $dx\,dy$ is replaced by $r\,dr\,d\theta$

Since $x = r\cos\theta$, $y = r\sin\theta$, we have $x^2 + y^2 = r^2$

Hence in polar coordinates,

$$\iint\limits_{D} e^{-(x^2+y^2)} dx\,dy = \int_{r=1}^{\infty} \int_{\theta=0}^{2\pi} e^{-r^2} r\, d\theta\, dr$$

$$= \int_{r=1}^{\infty} e^{-r^2} r\, [2\pi]\, dr = 2\pi \left(-\tfrac{1}{2} e^{-r^2}\right)\Big|_{1}^{\infty}$$

$$= 2\pi(0) - 2\pi\left(-\tfrac{1}{2} e^{-1}\right) = \boxed{\dfrac{\pi}{e}}$$

■■■ **4-41**

Evaluate the double integral

$$\iint_R e^{-(x^2 + y^2)} \, dA$$

Where R is the region in the first quadrant bounded by the circle $x^2 + y^2 = 1$ and the coordinate axes.

****

A sketch of the region of integration:

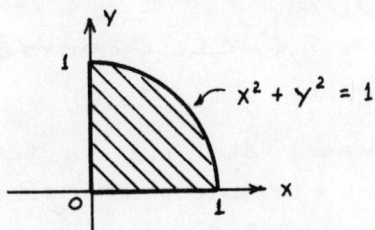

Since part of the boundary is circular, it will be helpful to use polar coordinates.

Note that $r^2 = x^2 + y^2$, $0 \leq r \leq 1$ and $0 \leq \theta \leq \pi/2$

So we have

$$\iint_R e^{-(x^2 + y^2)} \, dA = \int_0^{\pi/2} \int_0^1 e^{-r^2} r \, dr \, d\theta$$

$$= -\frac{1}{2} \int_0^{\pi/2} \int_0^{-1} e^u \, du \, d\theta$$

$$= -\frac{1}{2} (e^{-1} - 1) \int_0^{\pi/2} d\theta$$

$$= \frac{\pi}{4} (1 - e^{-1})$$

4-42

A paraboloid described by the equation $z = 1 - x^2 - y^2$ and the xy plane would create a volume. How would you ascertain this volume by using double integrals in polar coordinates ?

****

Let D be the whole unit disk in the xy plane, so that we write :

$$D = \left\{ (x,y) : x^2 + y^2 \leq 1 \right\}$$

and in polar coordinates we have :

$$D = \left\{ (r, \theta) : 0 \leq r \leq 1, 0 \leq \theta \leq 2\pi \right\}$$

Now, by using double integrals the required volume is given by :

$$V = \iint_D (1 - x^2 - y^2) \, dA$$

and in polar coordinates we have :

$$V = \int_0^{2\pi} \int_0^1 (1 - r^2) \, r \, dr \, d\theta$$

Evaluating the inside integral, we get

$$\int_0^1 (1 - r^2) r \, dr = \int_0^1 (r - r^3) \, dr$$

$$= \frac{r^2}{2} - \frac{r^4}{4} \Big/_0^1$$

$$= \left[\frac{(1)^2}{2} - \frac{(1)^4}{4} \right] - \left[\frac{(0)^2}{2} - \frac{(0)^4}{4} \right]$$

$$= \frac{1}{2} - \frac{1}{4} = \frac{1}{4}$$

Hence,

$$V = \int_0^{2\pi} \frac{1}{4} \, d\theta$$

$$= \frac{1}{4} \theta \Big/_0^{2\pi}$$

$$= \frac{1}{4} (2\pi - 0) = \frac{2\pi}{4} = \frac{\pi}{2} \quad \underline{Ans.}$$

■■■■■■■■■■■■■■■■■■■■■■■■■■■■■■■■■■■■■■■ **4-43**

Find the area inside the rose
$$r = 2a \sin 2\theta \text{ and outside the circle } r = a.$$

****

Solving simultaneously, $a = 2a \sin 2\theta$, $\sin 2\theta = \frac{1}{2}$, so in the first octant the curves intersect at $15°$ ($\pi/12$) and $75°$ ($5\pi/12$). By symmetry,

$$A = 4 \int_{\pi/12}^{5\pi/12} \int_{a}^{2a\sin 2\theta} r \, dr \, d\theta$$

$$= 2 \int_{\pi/12}^{5\pi/12} (2a \sin 2\theta)^2 - a^2 \, d\theta$$

$$= 2a^2 \int_{\pi/12}^{5\pi/12} 4\sin^2 2\theta - 1 \, d\theta$$

$$= 2a^2 \int_{\pi/12}^{5\pi/12} 2 - 2\cos 4\theta - 1 \, d\theta$$

$$= 2a^2 \left[\frac{\pi}{3} - \frac{1}{2} \sin 4\theta \Big|_{\pi/12}^{5\pi/12} \right]$$

$$= 2a^2 \left(\frac{\pi}{3} - \frac{1}{2} \sin \frac{5\pi}{3} + \frac{1}{2} \sin \frac{\pi}{3} \right)$$

$$= 2a^2 \left(\frac{\pi}{3} + \frac{\sqrt{3}}{2} \right) = \frac{a^2}{3} (2\pi + 3\sqrt{3}).$$

4-44 ▬▬▬▬▬▬▬▬▬▬▬▬▬▬▬▬▬▬▬▬▬▬▬

Evaluate $\displaystyle\int_{-2}^{2}\int_{0}^{\sqrt{4-x^2}}\sqrt{9-x^2-y^2}\ dy\ dx$

****

Change to polar

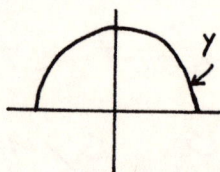

$y=\sqrt{4-x^2}$

$x^2+y^2=4=r^2 \rightarrow r=2$

$$\int_{-2}^{2}\int_{0}^{\sqrt{4-x^2}}\sqrt{9-(x^2+y^2)}\ dy\,dx = \int_{0}^{\pi}\int_{0}^{2}\sqrt{9-r^2}\ r\,dr\,d\theta$$

$$= -\frac{1}{2}\int_{0}^{\pi}\frac{2}{3}(9-r^2)^{3/2}\Big|_{0}^{2}d\theta = -\frac{1}{2}\int_{0}^{\pi}\left[\frac{2}{3}(5)^{3/2}-\frac{2}{3}(9)^{3/2}\right]d\theta$$

$$= \left[-\frac{1}{3}(5)^{3/2}+\frac{1}{3}(9)^{3/2}\right]\int_{0}^{\pi}d\theta = \left(9-\frac{1}{3}5^{3/2}\right)\theta\Big|_{0}^{\pi}$$

$$= \left(9-\frac{5^{3/2}}{3}\right)\pi$$

■■ **4-45**

Find the area of the region which lies inside the cardioid r = 5(1 + cos θ) and outside the circle r = 5.

****

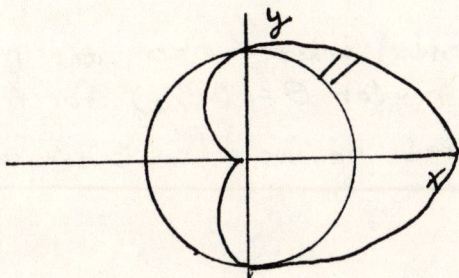

$$A = 2 \int_0^{\frac{\pi}{2}} d\theta \int_5^{5(1+\cos\theta)} r\, dr$$

$$= 2 \int_0^{\frac{\pi}{2}} \frac{r^2}{2} \Big|_5^{5(1+\cos\theta)} d\theta$$

$$= 25 \int_0^{\frac{\pi}{2}} (1 + 2\cos\theta + \cos^2\theta - 1)\, d\theta$$

$$= 25 \int_0^{\frac{\pi}{2}} \left(2\cos\theta + \frac{1}{2} + \frac{\cos 2\theta}{2} \right) d\theta$$

$$= 25 \left[2\sin\theta + \frac{1}{2}\theta + \frac{\sin 2\theta}{4} \right]_0^{\frac{\pi}{2}}$$

$$= 25 \left(2 + \frac{\pi}{4} \right)$$

4-46 ■■■

Find the area inside one leaf of the rose $r = 2 \sin 3\theta$, using double integrals.

****

A leaf begins (and ends) where $r = 0$, and $r = 2 \sin 3\theta = 0$ whenever $3\theta = n\pi$ (or $\theta = n\pi/3$) for integer n values. Therefore, one leaf begins when $\theta = 0$ and ends when $\theta = \pi/3$.

$$A = \int_0^{\pi/3} \int_0^{2\sin 3\theta} r \, dr \, d\theta = \int_0^{\pi/3} \left[\frac{r^2}{2}\right]_0^{2\sin 3\theta} d\theta$$

$$= \int_0^{\pi/3} \frac{4\sin^2 3\theta}{2} \, d\theta = \int_0^{\pi/3} 2\left(\frac{1-\cos 6\theta}{2}\right) d\theta$$

$$= \left[\theta - \frac{\sin 6\theta}{6}\right]_0^{\pi/3} = \left[\frac{\pi}{3} - \frac{\sin 2\pi}{6}\right] - 0 = \pi/3$$

CENTER OF MASS AND MOMENT OF INERTIA

---**4-47**

Let R be the region bounded by $y = x^2$, $y = 0$, $x = 1$. Find the center of mass of a lamina in the shape of R with density function $\rho(x,y) = x \cdot y$.

****

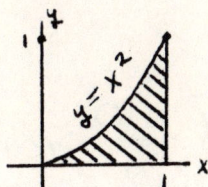

$$M = \int_0^1 \int_0^{x^2} xy \, dy \, dx = \int_0^1 x \frac{y^2}{2} \Big|_0^{x^2} dx = \frac{1}{2} \int_0^1 x^5 dx = \frac{1}{2} \frac{x^6}{6} \Big|_0^1 = \frac{1}{12}.$$

$$M_x = \int_0^1 \int_0^{x^2} xy^2 \, dy \, dx = \int_0^1 x \frac{y^3}{3} \Big|_0^{x^2} dx = \frac{1}{3} \int_0^1 x^7 dx = \frac{1}{3} \frac{x^8}{8} \Big|_0^1 = \frac{1}{24}.$$

$$M_y = \int_0^1 \int_0^{x^2} x^2 y \, dy \, dx = \int_0^1 x^2 \frac{y^2}{2} \Big|_0^{x^2} dx = \frac{1}{2} \int_0^1 x^6 dx = \frac{1}{2} \frac{x^7}{7} \Big|_0^1 = \frac{1}{14}.$$

$$\bar{x} = \frac{M_y}{M} = \frac{1}{14} \cdot \frac{12}{1} = \frac{6}{7}, \quad \bar{y} = \frac{M_x}{M} = \frac{1}{24} \cdot \frac{12}{1} = \frac{1}{2}, \quad (\bar{x}, \bar{y}) = \left(\frac{6}{7}, \frac{1}{2}\right).$$

4-48 ▬▬▬▬▬▬▬▬▬▬▬▬▬▬▬▬▬▬▬▬▬▬▬▬▬▬▬▬

Find the centroid of the plane region inside the cardoid r = 4(1+cosθ) and outside the circle r = 4.

****

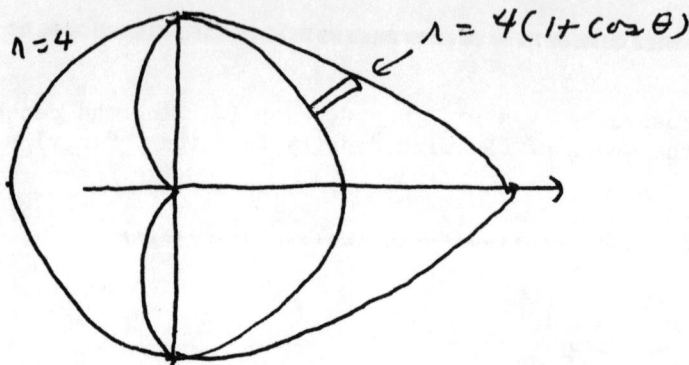

By Symmetry $\bar{y} = 0$

Area of Region $= 2 \int_0^{\pi/2} \left(\int_4^{4(1+\cos\theta)} (r \, dr) \right) d\theta$

$$= 2 \int_0^{\pi/2} \frac{1}{2} r^2 \Big|_4^{4(1+\cos\theta)} d\theta$$

$$= \int_0^{\pi/2} 16(1+\cos\theta)^2 - 16 \; d\theta$$

$$= \int_0^{\pi/2} (32\cos\theta + 16\cos^2\theta) \, d\theta$$

$$= \int_0^{\pi/2} 32\cos\theta + 8(1+\cos 2\theta) \, d\theta$$

$$= 32\sin\theta + 8\theta + 4\sin 2\theta \Big|_0^{\pi/2}$$

$$= 32 + 4\pi + 0 - (0+0+0) = 32 + 4\pi$$

Now $\bar{x} = \dfrac{M_y}{A}$; $M_y = \iint x \, dA.$

$$M_y = 2\int_0^{\pi/2} d\theta \int_4^{4(1+\cos\theta)} \wedge^2 \cos\theta \, dA.$$

$$= 2\int_0^{\pi/2} \cos\theta \left[\frac{\wedge^3}{3} \bigg|_4^{4(1+\cos\theta)} \right] d\theta$$

$$= \frac{2}{3}\int_0^{\pi/2} \cos\theta \left[64 + 192\cos\theta + 192\cos^2\theta + 64\cos^3\theta - 64 \right] d\theta.$$

$$= \frac{128}{3}\int_0^{\pi/2} 3\cos^2\theta + 3\cos^3\theta \, d\theta$$

$$= 128\int_0^{\pi/2} \left[\frac{1}{2} + \frac{1}{2}\cos 2\theta \right] + \cos\theta(1-\sin^2\theta) d\theta$$

$$= 128\left(\frac{\theta}{2} + \frac{1}{4}\sin 2\theta + \sin\theta - \frac{\sin^3\theta}{3} \bigg|_0^{\pi/2} \right)$$

$$= 128\left[\left(\frac{\pi}{4} + 1 - \frac{1}{3} \right) - \left(0 + 0 + 0 - 0 \right) \right]$$

$$= 32\pi + \frac{256}{3} = \frac{96\pi + 212}{3}$$

$$\text{So } \overline{x} = \frac{(96\pi + 202)}{96 + 12\pi}$$

4-49 ■■

Find the moment of inertia with respect to its axis of a right circular cone of base radius R and height H.

****

Consider the given cone as the line $y = \frac{H}{R}x$ rotated about the z-axis.

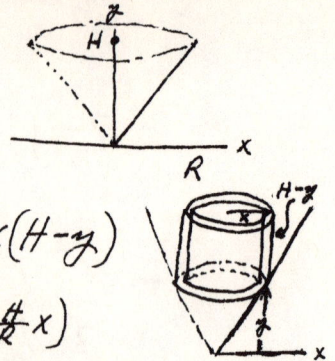

$$I = \int r^2 dm \qquad dm = \rho \, dV = \rho \, 2\pi x \, dx(H-y)$$

Since $y = \frac{H}{R}x$, $\quad dm = \rho 2\pi x \, dx \left(H - \frac{H}{R}x\right)$

$$I = \int_0^R x^2 \rho 2\pi x \, dx \left(H - \frac{H}{R}x\right) = 2\pi \rho H \int_0^R x^3 \left(1 - \frac{x}{R}\right) dx$$

$$= 2\pi \rho H \int_0^R \left(x^3 - \frac{x^4}{R}\right) dx = 2\pi \rho H \left(\frac{x^4}{4} - \frac{x^5}{5R}\right)\Big|_0^R$$

$$= 2\pi \rho H \left(\frac{R^4}{4} - \frac{R^5}{5R} - 0\right) = 2\pi \rho H \left(\frac{5R^4 - 4R^4}{20}\right)$$

$$= 2\pi \rho H \frac{R^4}{20} = \pi \rho H \frac{R^4}{10}$$

The mass of the cone $= M = \rho V = \rho \frac{1}{3}\pi R^2 H$

$$I = \pi \rho H \frac{R^4}{10} = \frac{R^2}{10} 3 \cdot \frac{1}{3}\pi R^2 H = \frac{R^2}{10} 3M$$

Thus, $\quad I = \frac{3}{10} M R^2$.

■■■**4-50**

Find the moment of inertia, I_x , about the x-axis and the moment of inertia, I_y , about the y-axis for the region in the first quadrant bounded by y= x and $y^2 = x^3$.

****

The region is shown in the following diagram:

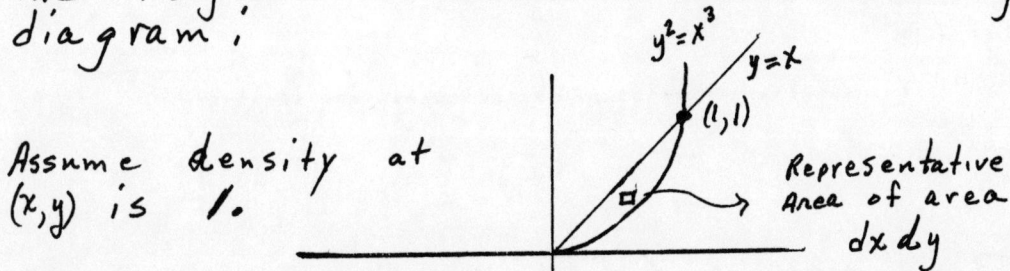

Assume density at (x,y) is 1.

The representative area is y units from the x-axis; hence, by I_x we have

$$I_x = \int_0^1 \int_{x_1=y}^{x_2=y^{\frac{2}{3}}} y^2 \, dx \, dy = \int_0^1 \left(y^{\frac{8}{3}} - y^3 \right) dy$$

$$= \left(\frac{3}{11} y^{\frac{11}{3}} - \frac{y^4}{4} \right)_0^1$$

$$= \frac{1}{44}$$

And

The representative area, $dx\,dy$, is x units from the y axis; hence, by the definition of I_y we have

$$I_y = \int_0^1 \int_{x_1=y}^{x_2=y^{\frac{2}{3}}} x^2 \, dx \, dy = \frac{1}{3} \int_0^1 \left(y^2 - y^3 \right) dy$$

$$= \frac{1}{3} \left[\frac{y^3}{3} - \frac{y^4}{4} \right]_0^1$$

$$= \frac{1}{36}$$

AREA OF A SURFACE

4-51 ■■

Find the surface area on the surface z = xy inside the cylinder
$x^2 + y^2 = 1$.

****

The integrand for surface area is

$$\sqrt{1+\left(\frac{\partial z}{\partial x}\right)^2+\left(\frac{\partial z}{\partial y}\right)^2} = \sqrt{1+y^2+x^2}.$$

Using Polar Coordinates,

$$A = \int_0^{2\pi}\int_0^1 r\sqrt{1+r^2}\,dr\,d\theta$$

$$= (\text{let } u = 1+r^2, du = 2r\,dr)$$

$$\int_0^{2\pi}\int_1^2 \frac{1}{2}\sqrt{u}\,du\,d\theta$$

$$= \int_0^{2\pi} \frac{1}{3}u^{3/2}\Big|_1^2\,d\theta$$

$$= \frac{2\pi}{3}\left(2\sqrt{2}-1\right).$$

■■ **4-52**

Compute the area of that part of the
graph of
$$3z = 5 + 2x^{3/2} + 4y^{3/2}$$

which lies above the rectangular
region in the first quadrant of
the xy-plane bounded by the lines
x = 0, x = 3, y = 0, and y = 6.

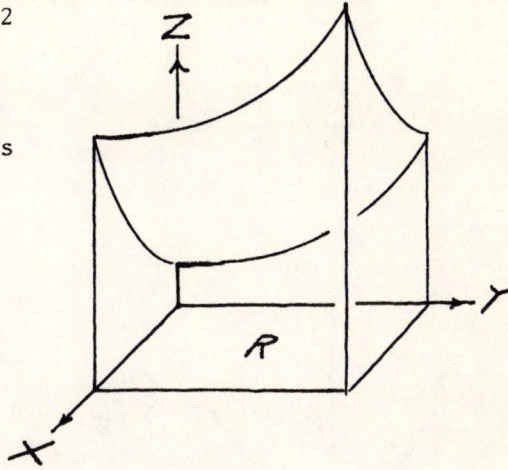

****

$$z = \frac{5}{3} + \frac{2}{3}x^{3/2} + \frac{4}{3}y^{3/2}, \quad \frac{\partial z}{\partial x} = x^{1/2}, \quad \frac{\partial z}{\partial y} = 2y^{1/2}$$

$$A = \iint_R \sqrt{\left(\frac{\partial z}{\partial x}\right)^2 + \left(\frac{\partial z}{\partial y}\right)^2 + 1}\; dA = \int_0^6 \int_0^3 \sqrt{x + 4y + 1}\; dx\, dy$$

$$= \int_0^6 \frac{2}{3}\Big[(x + 4y + 1)^{3/2}\Big]_0^3\; dy$$

$$= \frac{2}{3}\int_0^6 \Big[8(y+1)^{3/2} - (4y+1)^{3/2}\Big]\, dy$$

$$= \frac{2}{3}\Big[8\cdot\frac{2}{5}(y+1)^{5/2} - \frac{1}{4}\cdot\frac{2}{5}(4y+1)^{5/2}\Big]_0^6$$

$$= \frac{2}{3}\Big[\frac{16}{5}(7^{5/2} - 1) - \frac{1}{10}(5^5 - 1)\Big] = \frac{4}{15}(392\sqrt{7} - 789)$$

4-53 ▪▪▪▪▪▪▪▪▪▪▪▪▪▪▪▪▪▪▪▪▪▪▪▪▪▪▪▪▪▪▪▪

Find the surface area of the part of the cone $z = (x^2 + y^2)^{\frac{1}{2}}$ lying inside the cylinder $x^2 - 2x + y^2 = 0$.

****

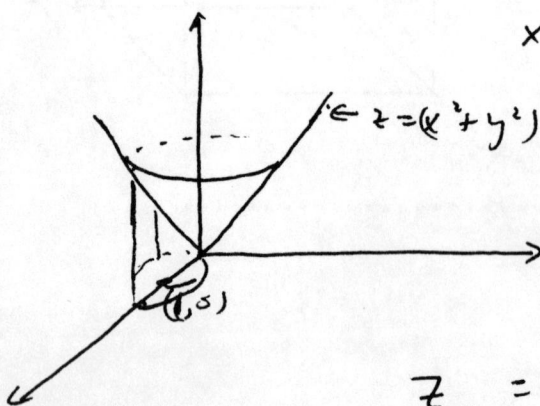

$x^2 - 2x + y^2 = 0$

so

$x^2 - 2x + 1 + y^2 = 1$

$(x-1)^2 + y^2 = 1.$

$z_x = 2x(x^2+y^2)^{-\frac{1}{2}}$

$z_y = 2y(x^2+y^2)^{-\frac{1}{2}}$

So $z_x^2 + z_y^2 + 1 = \dfrac{4x^2}{x^2+y^2} + \dfrac{4y^2}{x^2+y^2} + 1$

$= \dfrac{5x^2 + 5y^2}{x^2+y^2} = 5$

Thus

$Area = \iint_G 5\,dx\,dy = 5\,area\,G$

But G is circle radius 1 centered at $(1,0,0)$

So area $G = \pi$.

Thus 5π is surface area.

■■■-- **4-54**

Set up, but do not evaluate, the integral to find the surface area of the portion of the sphere $x^2 + y^2 + z^2 = 16$ between the planes z=1 and z=2.

****

$$SA = \int_R \sqrt{f_x^2 + f_y^2 + 1} \; dA$$

$$f(x,y) = \sqrt{16 - x^2 - y^2}$$

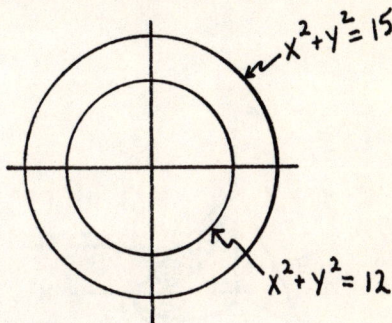

$$f_x = \frac{-x}{\sqrt{16 - x^2 - y^2}} \qquad f_y = \frac{-y}{\sqrt{16 - x^2 - y^2}}$$

$$SA = \int_R \sqrt{\frac{x^2}{16 - x^2 - y^2} + \frac{y^2}{16 - x^2 - y^2} + 1} \; dA$$

$$= \int_R \sqrt{\frac{16}{16 - x^2 - y^2}} \; dA = \int_R \sqrt{\frac{16}{16 - r^2}} \; dA$$

$$= \int_0^{2\pi} \int_{\sqrt{12}}^{\sqrt{15}} \sqrt{\frac{16}{16 - r^2}} \; r \, dr \, d\theta$$

(figure: annulus with outer circle labeled $x^2 + y^2 = 15$ and inner circle labeled $x^2 + y^2 = 12$)

4-55

Find the area of the surface cut from the cone $z = 1 - \sqrt{x^2 + y^2}$ by the cylinder $x^2 + y^2 = y$.

****

½ of required area

Region of Integration

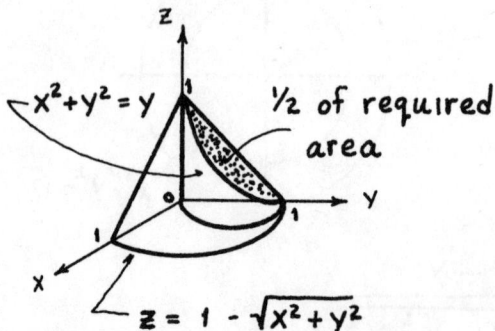

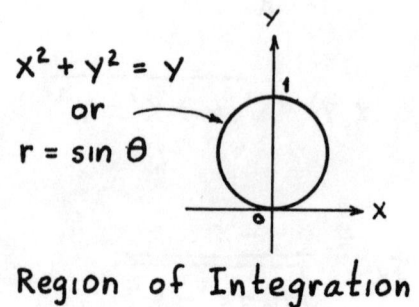

Begin by sketching the portion of the surface in the first octant. Note that by symmetry we may double the area in the first octant. It will also be helpful to use cylindrical coordinates.

Recall that surface area is given by:

$$S = \iint_R \sqrt{\left(\frac{\partial f}{\partial x}\right)^2 + \left(\frac{\partial f}{\partial y}\right)^2 + 1} \; dy\,dx$$

In cylindrical coordinates, $dy\,dx$ becomes $r\,dr\,d\theta$.

Now, for $f(x,y) = 1 - \sqrt{x^2 + y^2}$ we have

$$\frac{\partial f}{\partial x} = \frac{-x}{(x^2 + y^2)^{1/2}} \;\Rightarrow\; \left[\frac{\partial f}{\partial x}\right]^2 = \frac{x^2}{x^2 + y^2}$$

and

$$\frac{\partial f}{\partial y} = \frac{-y}{(x^2 + y^2)^{1/2}} \;\Rightarrow\; \left[\frac{\partial f}{\partial y}\right]^2 = \frac{y^2}{x^2 + y^2}$$

So $\left(\dfrac{\partial f}{\partial x}\right)^2 + \left(\dfrac{\partial f}{\partial y}\right)^2 + 1 = \dfrac{x^2}{x^2+y^2} + \dfrac{y^2}{x^2+y^2} + 1$

$$= \dfrac{2\,(x^2+y^2)}{x^2+y^2} = 2$$

Thus $\quad S = 2 \displaystyle\int_0^{\pi/2} \int_0^{\sin\theta} \sqrt{2}\; r\, dr\, d\theta$

$$= \sqrt{2} \int_0^{\pi/2} r^2 \Big]_0^{\sin\theta} d\theta$$

$$= \sqrt{2} \int_0^{\pi/2} \sin^2\theta\; d\theta$$

$$= \sqrt{2} \left(\dfrac{\theta}{2} - \dfrac{\sin 2\theta}{4}\right)\Big]_0^{\pi/2} \quad = \dfrac{\pi\sqrt{2}}{4}$$

4-56 ━━━

Find the portion of the surface $z = x^2 + y^2$ inside the cylinder $x^2 + y^2 = 4$.

****

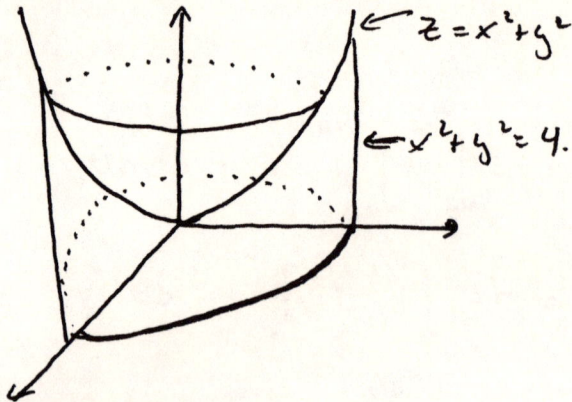

$$dS = \sqrt{\left(\frac{\partial z}{\partial x}\right)^2 + \left(\frac{\partial z}{\partial y}\right)^2 + 1} = \sqrt{4x^2 + 4y^2 + 1}$$

The Surface Lies Over The Disk $x^2 + y^2 = 4$.

Using Polar Coordinates $x = \Lambda \cos\theta$
$$dS = (1 + 4\Lambda^2)^{1/2}$$ $y = \Lambda \sin\theta$

and

$$SA = \int_0^{2\pi} \left(\int_0^2 (1 + 4\Lambda^2)^{1/2} \Lambda \, d\Lambda \right) d\theta$$

$u = 1 + 4\Lambda^2$ $\Lambda = 0$ $u = 1$.
$du = 8\Lambda \, d\Lambda$. $\Lambda = 2$ $u = 17$

$$= \frac{2\pi}{8} \int_1^{17} u^{1/2} du = \frac{\pi}{4} \left(\frac{2}{3} u^{3/2} \Big|_1^{17} \right)$$

$$= \frac{\pi}{6} (17\sqrt{17} - 1)$$

GREEN'S THEOREM

■■■ **4-57**

Verify Green's Theorem for $\int_C M\,dx + N\,dy$ where $M = xy$, $N = x^2$, C the boundary curve of the triangle with vertices $(0,0)$, $(1,0)$, and $(1,1)$.

****

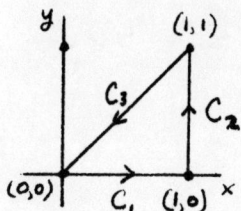

Green's theorem says $\int_C M\,dx + N\,dy = \iint_R (N_x - M_y)\,dA$, where R is the region inside C. $C = C_1 + C_2 + C_3$,

C_1: $x = t$, $y = 0$, $0 \le t \le 1$, $\int_{C_1} xy\,dx + x^2\,dy =$

$\int_0^1 (0 \cdot 1 + t^2 \cdot 0)\,dt = 0$. C_2: $x = 1$, $y = t$, $0 \le t \le 1$, $\int_{C_2} xy\,dx + x^2\,dy$

$= \int_0^1 (1 \cdot t \cdot 0 + 1^2 \cdot 1)\,dt = \int_0^1 dt = 1$. C_3: $x = 1-t$, $y = 1-t$, $0 \le t \le 1$,

$\int_{C_3} xy\,dx + x^2\,dy = \int_0^1 \left[(1-t)(1-t)(-1) + (1-t)^2(-1)\right]dt = -2\int_0^1 (1-t)^2\,dt$

$= -2 \left.\frac{(1-t)^3}{-3}\right|_0^1 = -\frac{2}{3}$. So $\int_C xy\,dx + x^2\,dy = 0 + 1 - \frac{2}{3} = \frac{1}{3}$.

Now $\iint_R (N_x - M_y)\,dA = \int_0^1 \int_0^x (2x - x)\,dy\,dx = \int_0^1 \int_0^x x\,dy\,dx$

$= \int_0^1 x^2\,dx = \left.\frac{x^3}{3}\right|_0^1 = \frac{1}{3}$ also.

4-58

Evaluate the line integral $\oint xy^2dx - yx^2dy$ around the triangle with vertices (1,0), (0,1), and (0,0) with clockwise orientation.

****

RECALL GREENS THM

$$\oint_C Pdx + Qdy = \iint_G \left(\frac{\partial Q}{\partial x} - \frac{\partial P}{\partial y} \right) dxdy.$$

$G = $ int C.

$$\frac{\partial Q}{\partial x} = -2xy \left. \right\} \text{ so } \frac{\partial Q}{\partial x} - \frac{\partial P}{\partial y} = -4xy$$

$$\frac{\partial P}{\partial y} = 2xy$$

G:

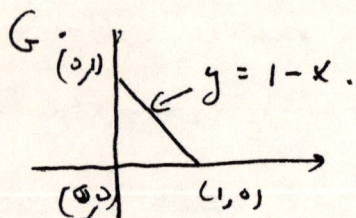

So

$$I = -4\int_0^1 \left(\int_0^{(1-x)} xy\, dy \right) dx = -4\int_0^1 x \left(\frac{y^2}{2} \Big|_0^{1-x} \right) dx$$

$$= -4\int_0^1 \frac{x(1-x)^2}{2}dx = -2\int_0^1 (x - 2x^2 + x^3)dx$$

$$= -2\left[\frac{x^2}{2} - \frac{2}{3}x^3 + \frac{x^4}{4} \Big|_0^1 \right] = -2\left(\frac{1}{2} - \frac{2}{3} + \frac{1}{4} \right)$$

$$= -2\left(\frac{1}{12} \right) = -\frac{1}{6}.$$

■■■4-59

Let C be the closed path from (0,0) to (2,4) along $y = x^2$ and back again from (2,4) to (0,0) along $y = 2x$. Evaluate

$$\int_C (x^3+2y)\,dx + (x-y^2)\,dy$$

(a) directly
(b) using Green's theorem.

****

(a) $\displaystyle\int_C (x^3+2y)\,dx + (x-y^2)\,dy$

$\displaystyle = \int_0^2 (x^3+2x^2)\,dx + (x-x^4)(2x)\,dx$

$\displaystyle\quad + \int_2^0 (x^3+4x)\,dx + (x-4x^2)(2)\,dx$

$\displaystyle = \int_0^2 (-2x^5+x^3+4x^2)\,dx + \int_2^0 (x^3-8x^2+6x)\,dx$

$\displaystyle = -\frac{2x^6}{6} + \frac{x^4}{4} + \frac{4x^3}{3}\Big|_0^2 + \frac{x^4}{4} - \frac{8x^3}{3} + 3x^2\Big|_2^0$

$\displaystyle = -\frac{64}{3} + 4 + \frac{32}{3} - 4 + \frac{64}{3} - 12 = -4/3$

(b) $P(x,y) = x^3+2y$, $Q(x,y) = x-y^2$

$\displaystyle\int_C P\,dx + Q\,dy = \int_0^2\int_{x^2}^{2x}\left[\frac{\partial Q}{\partial x} - \frac{\partial P}{\partial y}\right]dy\,dx$

$\displaystyle = \int_0^2\int_{x^2}^{2x}(1-2)\,dy\,dx = -\int_0^2 (2x-x^2)\,dx$

$\displaystyle = -\left[x^2 - \frac{x^3}{3}\right]_0^2 = -\left[4 - 8/3\right] = -4/3$

4-60

If R is the region enclosed by a simple closed positively oriented curve C, then the area of R is given by

(a) $\int_C y\ dx + 2x\ dy$

(b) $\int_C x\,dy + y\,dx$

(c) $1/2 \int_C y\ dx + x\ dy$

(d) $1/2 \int_C y\ dx - x\ dy$

(e) none of (a) - (d) are correct.

****

In (a), $M = y$, $N = 2x$, $N_x - M_y = 2 - 1 = 1$, so by Green's theorem, $\int_C y\,dx + 2x\,dy = \iint_R (N_x - M_y)\,dA = \iint_R dA = $ area of R.

━━━━━━━━━━━━━━━━━━━━━━━━━━━━━━━━━━━**4-61**

Evaluate the following line integral.

$$\int_c (10x^4 - 2xy^3)\,dx - 3x^2y^2\,dy$$

where c is the curve along the path $x^4 - 6xy^3 = 4y^2$ from (0,0) to (2,1).

****

Setting $M(x,y) = 10x^4 - 2xy^3$ and $N(x,y) = -3x^2y^2$

we see $M_y = -6xy^2 = N_x$

Thus is $\varphi(x,y)$ is a potential,

$$\int_c M\,dx + N\,dy = \varphi(2,1) - \varphi(0,0)$$

$$\varphi(x,y) = \int 10x^4 - 2y^3x\,dx \quad \text{and} \quad \varphi(x,y) = \int -3x^2y^2\,dy$$

$$= 2x^5 - x^2y^3 + f(y) \qquad\qquad = -x^2y^3 + g(x)$$

thus $\varphi(x,y) = 2x^5 - x^2y^3 + c$

$\therefore \displaystyle\int_c M\,dx + N\,dy = \varphi(2,1) - \varphi(0,0)$

$$= (64 - 4) - 0$$

$$= \boxed{60}$$

Note the alternative would have been to parameterize the curve, not a pretty prospect!

THE TRIPLE INTEGRAL
RECTANGULAR COORDINATES

4-62 ■■■

Find $\iiint_R 2x \, dV$ where R is the region in the first octant bounded by the cylinders $z = 1-y^2$ and $z = 1-x^2$.

****

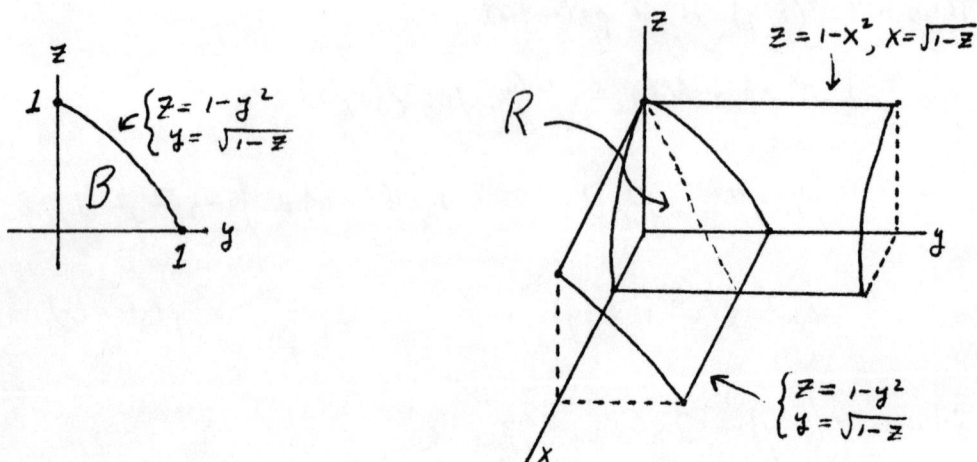

If we view R as having its base in the yz plane, then R is bounded below by $x=0$, above by $x=\sqrt{1-z}$, and laterally by the cylinder which intersects the yz plane in the region B shown above. Therefore,

$$\iiint_R 2x \, dV = \iint_B \left[\int_{x=0}^{x=\sqrt{1-z}} 2x \, dx \right] dA = \int_{z=0}^{z=1} \int_{y=0}^{y=\sqrt{1-z}} \int_{x=0}^{x=\sqrt{1-z}} 2x \, dx \, dy \, dz$$

$$= \int_0^1 \int_0^{\sqrt{1-z}} x^2 \Big|_0^{\sqrt{1-z}} \, dy \, dz = \int_0^1 \int_0^{\sqrt{1-z}} (1-z) \, dy \, dz = \int_0^1 (1-z) y \Big|_{y=0}^{y=\sqrt{1-z}} \, dz$$

$$= \int_0^1 (1-z)^{\frac{3}{2}} \, dz = -\frac{2}{5} (1-z)^{\frac{5}{2}} \Big|_0^1 = \left(-\frac{2}{5}\right) \cdot (0-1) = \frac{2}{5}.$$

4-63

Find the volume, using triple integrals, of the region in the first octant beneath the plane $x + 2y + 3z = 6$.

****

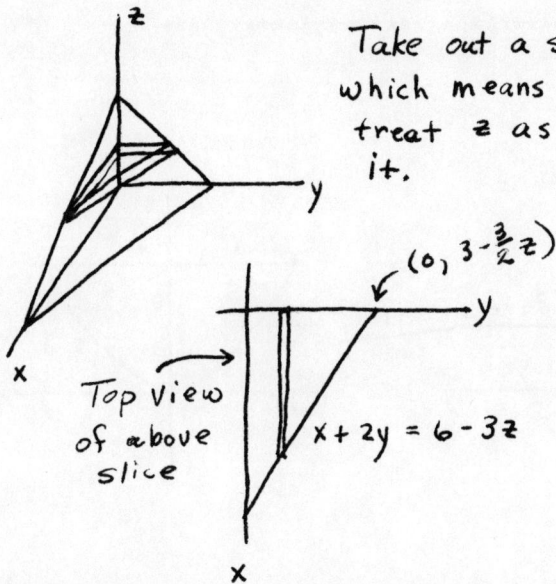

Take out a slice parallel to the xy plane, which means we integrate z last, and we treat z as a constant until we integrate it.

$(0, 3 - \frac{3}{2}z)$

Top view of above slice

$x + 2y = 6 - 3z$

Take out a rectangle parallel to the x axis, which means we integrate y second to last, and we hold y constant until we integrate it.

$$V = \int_0^2 \int_0^{3 - \frac{3}{2}z} \int_0^{6 - 3z - 2y} 1 \, dx \, dy \, dz = \int_0^2 \int_0^{3 - \frac{3}{2}z} \Big[x \Big]_0^{6 - 3z - 2y} dy \, dz$$

$$= \int_0^2 \int_0^{3 - \frac{3}{2}z} (6 - 3z - 2y) \, dy \, dz = \int_0^2 \Big(6y - 3yz - y^2 \Big)_0^{3 - \frac{3}{2}z} dz$$

$$= \int_0^2 \Big(6(3 - \frac{3}{2}z) - 3z(3 - \frac{3}{2}z) - (3 - \frac{3}{2}z)^2 \Big) \, dz = \int_0^2 \Big(9 - 9z + \frac{9}{4}z^2 \Big) dz$$

$$= \Big[9z - \frac{9}{2}z^2 + \frac{3}{4}z^3 \Big]_0^2 = 9(2) - \frac{9}{2}(4) + \frac{3}{4}(8) - 0$$

$$= 18 - 18 + 6 = 6$$

4-64

Find $\iiint_S x^2y \, dV$ where S is the solid bounded by the cylinder $y = x^2$ and the planes $z = 0$, $y = 1$, $z = y$.

****

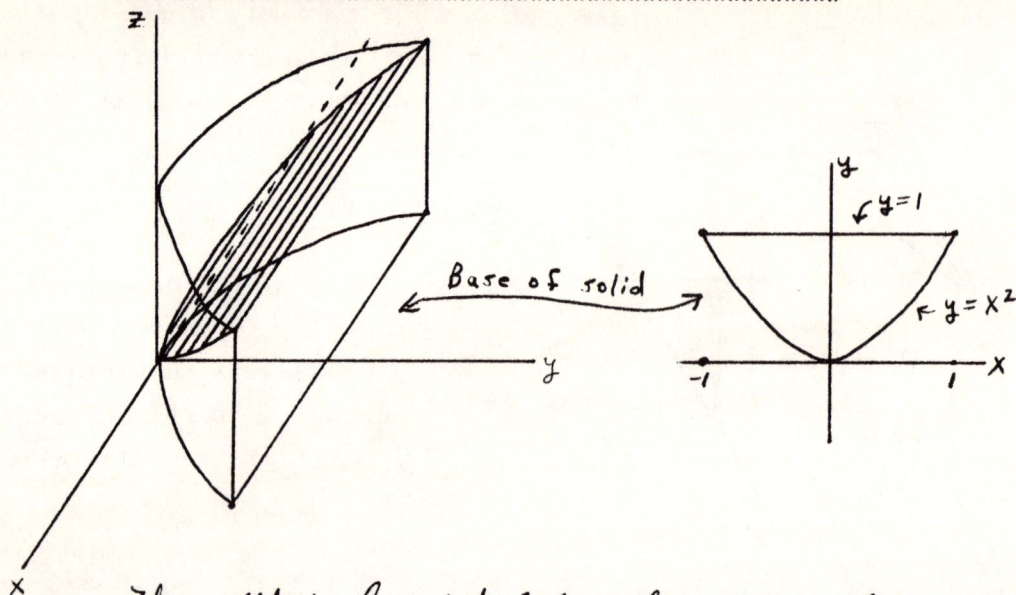

Base of solid

The solid is bounded below by $z = 0$, above by $z = y$, and laterally by the cylinder whose intersection with the xy plane is the base of the solid.

Hence $\iiint_S x^2y \, dV = \int_{x=-1}^{x=1} \int_{y=x^2}^{y=1} \int_{z=0}^{z=y} x^2y \, dz \, dy \, dx =$

$\int_{-1}^{1} \int_{x^2}^{1} x^2y^2 \, dy \, dx = \int_{-1}^{1} x^2 \frac{y^3}{3} \Big|_{x^2}^{1} dx = \frac{1}{3} \int_{-1}^{1} x^2(1 - x^6) \, dx$

$= \frac{1}{3} \left[\frac{x^3}{3} - \frac{x^9}{9} \right]_{-1}^{1} = \frac{4}{27}.$

4-65

Given that

$$V = \int_0^3 \int_0^{\frac{1}{2}(3-z)} \int_0^{4-x^2} dy \, dx \, dz$$

a) Sketch the solid whose volume is given by V.

b) Evaluate the integral to find the volume of the solid.

****

a) Note that the integration occurs over a region in the xz-plane and that

   y goes from 0 to 4-x²
   (parabola in the xy-plane)

   x goes from 0 to ½(3-z)
   (line in the xz-plane)

   z goes from 0 to 3

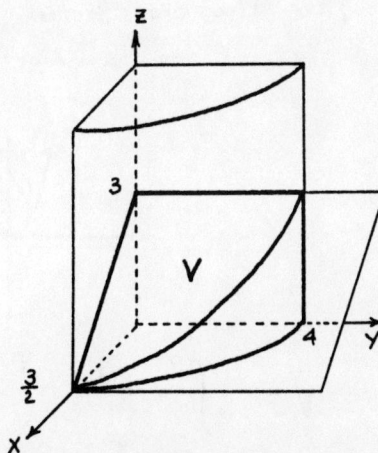

b) $V = \int_0^3 \int_0^{\frac{1}{2}(3-z)} (4 - x^2) \, dx \, dz$

$= \int_0^3 \left[6 - 2z - \frac{(3-z)^3}{24} \right] dz$

$= \left(6z - z^2 + \frac{(3-z)^4}{96} \right) \Big]_0^3$

$= (18 - 9 + 0) - (0 - 0 + {3^4}/{96})$

$= \dfrac{261}{32}$

4-66 ■■

Find the volume of the solid formed by the intersection of the cylinder
$$y = x^2$$
and the two planes given by the equations
$$z = 0$$
$$y + z = 4.$$

****

The planes $y + z = 4$ and $z = 0$ intersect in a line where $y = 4$ and $z = 0$

Thus the cross section of the solid where $z = 0$ is that part of the interior of the parabola $y = x^2$ where $y \leq 4$:

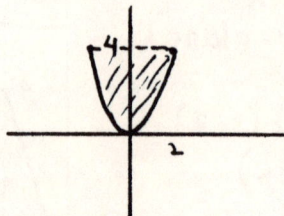

The solid consists of this area as a base, with elements "going up" to the plane $y + z = 4$

$$V = \int_{-2}^{2} \int_{x^2}^{4} \int_{0}^{4-y} dz\, dy\, dx$$

$$= \int_{-2}^{2} \int_{x^2}^{4} (4-y)\, dy\, dx$$

$$= \int_{-2}^{2} \left[4y - \frac{y^2}{2} \Big|_{x^2}^{4} \right] dx$$

$$= \int_{-2}^{2} 8 - 4x^2 + \frac{x^4}{4}\, dx$$

$$= 8x - \frac{4}{3}x^3 + \frac{x^5}{5} \Big|_{-2}^{2}$$

$$= \boxed{\frac{256}{15}}$$

━━━4-67

Evaluate $\displaystyle\int_0^1 \int_z^{2z} \int_{x+z}^{2x+2z} (x-1)\, dy\, dx\, dz$

****

$$\int_0^1 \int_z^{2z} \int_{x+z}^{2x+2z} (x-1)\, dy\, dx\, dz$$

$$= \int_0^1 \int_z^{2z} \Big[xy - y \Big]_{x+z}^{2x+2z} \, dx\, dz$$

$$= \int_0^1 \int_z^{2z} \left(2x^2 + 2xz - 2z - 2x - x^2 - xz + x + z \right) dx\, dz$$

$$= \int_0^1 \int_z^{2z} \left(x^2 + xz - x - z \right) dx\, dz$$

$$= \int_0^1 \left[\frac{x^3}{3} + \frac{x^2 z}{2} - \frac{x^2}{2} - xz \right]_z^{2z} dz$$

$$= \int_0^1 \left(\frac{8z^3}{3} + 2z^3 - 2z^2 - 2z^2 - \frac{z^3}{3} - \frac{z^3}{2} + \frac{z^2}{2} + z^2 \right) dz$$

$$= \int_0^1 \left(\frac{23}{6} z^3 - \frac{5}{2} z^2 \right) dz$$

$$= \left[\frac{23}{24} z^4 - \frac{5}{6} z^3 \right]_0^1 = \frac{23}{24} - \frac{5}{6} = \frac{1}{8}$$

4-68

Use the method of iterated integration in order to evaluate the triple

integral $\iiint\limits_{N} x \, dV$, where N is a region cut off from the first octant

by the plane defined by x + y + z = 3.

****

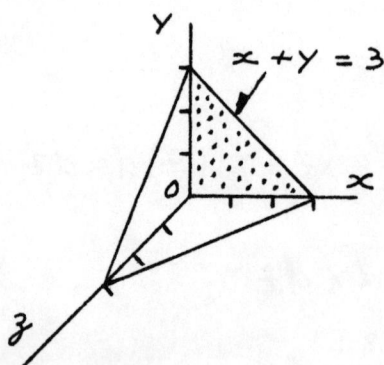

A view of the region is shown in the accompaning illustration. In order to carry out the intended integration let us do the first integration with respect to z, i.e. hold x and y fixed. By re-writing the given equation we would then have z varying from $z = 0$ to $z = 3 - x - y$, and our first integral would be:

$$\int_{0}^{3-x-y} x \, dz = xz \bigg|_{z=0}^{z=3-x-y}$$

$$= x(3-x-y) - x(0) = x(3-x-y)$$

We now have to deal with a double integral involving x and y, whose region is the shaded triangle shown in the illustration with the line $x + y = 3$.

Integrating now with respect to y (as a matter of choice), we have $y = 3 - x$ and write

$$\int_{0}^{3} \int_{0}^{3-x} x(3-x-y) \, dy \, dx$$

$$= \int_{0}^{3} \left[-x \frac{(3-x-y)^2}{2} \right]_{0}^{3-x} dx$$

$$= \frac{1}{2} \int_{0}^{3} x(3-x)^2 \, dx$$

$$= \frac{1}{2} \int_0^3 \left(9x - 6x^2 + x^3\right) dx$$

$$= \frac{1}{2} \left[\frac{9}{2} x^2 - \frac{6}{3} x^3 + \frac{x^4}{4} \right]_0^3$$

$$= \frac{1}{2} \left[\frac{9(3)^2}{2} - \frac{6(3)^3}{3} + \frac{(3)^4}{4} \right]$$
$$- \left[\frac{9(0)^2}{2} - \frac{6(0)^3}{3} + \frac{(0)^4}{4} \right]$$

$$= \frac{1}{2} \left[\frac{81}{2} - 54 + \frac{81}{4} - 0 + 0 - 0 \right]$$

$$= \frac{1}{2} \cdot \frac{27}{4} \qquad = \frac{27}{8} \qquad \underline{Ans.}$$

------------------------------- **4-69**

Evaluate the following multiple integral

$$\int_0^2 \int_0^{2z} \int_0^{xz} xyz^2 \, dy \, dx \, dz$$

****

$$\int_0^2 \int_0^{2z} \int_0^{x} xyz^2 \, dy \, dx \, dz = \int_0^2 \int_0^{2z} \frac{xy^2z^2}{2} \Big|_0^{xz} \, dx \, dz$$

$$= \int_0^2 \int_0^{2z} \frac{x^3 z^4}{2} \, dx \, dz$$

$$= \int_0^2 \frac{x^4 z^4}{8} \Big|_0^{2z} \, dz = \int_0^2 2z^8 \, dz$$

$$= \frac{2}{9} z^9 \Big|_0^2 = \frac{1024}{9}$$

4-70 ▪▪▪

Evaluate $\int_0^1 \int_{x^2}^1 \int_0^{3y} (y + 2x^2 z)\ dz\ dy\ dx$.

****

$$\int_0^{3y} (y + 2x^2 z)\ dz = (yz + x^2 z^2)\Big|_{z=0}^{z=3y}$$

$$= 3y^2 + 9x^2 y^2$$

$$\int_{x^2}^1 (3y^2 + 9x^2 y^2)\ dy = (y^3 + 3x^2 y^3)\Big|_{y=x^2}^{y=1}$$

$$= (1 + 3x^2) - (x^6 + 3x^8) = 1 + 3x^2 - x^6 - 3x^8$$

$$\int_0^1 (1 + 3x^2 - x^6 - 3x^8)\ dx = \left(x + x^3 - \frac{x^7}{7} - \frac{x^9}{3}\right)\Big|_0^1$$

$$= 1 + 1 - \frac{1}{7} - \frac{1}{3} = \frac{32}{21}$$

THE TRIPLE INTEGRAL
CYLINDRICAL COORDINATES

■■ **4-71**

Use cylindrical coordinates to find $\iiint_R z \, dV$ where R is the region bounded by $z = \sqrt{x^2 + y^2}$ and $z = x^2 + y^2$.

****

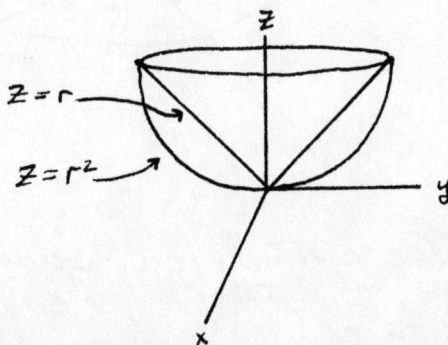

The solid is bounded above by the cone $z = \sqrt{x^2+y^2}$, ie. $z = r$, and below by the paraboloid $z = x^2+y^2$, ie. $z = r^2$. Solving these simultaneously gives $r=1$, so the solid is bounded laterally by the cylinder $r=1$ which intersects the xy plane in the unit circle $r=1$.

Therefore, $\iiint_R z \, dV = \int_0^{2\pi} \int_0^1 \int_{r^2}^{r} z \cdot r \, dz \, dr \, d\theta$

$$= \int_0^{2\pi} \int_0^1 r \frac{z^2}{2} \Big|_{r^2}^{r} dr \, d\theta = \frac{1}{2} \int_0^{2\pi} \int_0^1 (r^3 - r^5) \, dr \, d\theta$$

$$= \frac{1}{2} \int_0^{2\pi} \left[\frac{r^4}{4} - \frac{r^5}{5} \right]_0^1 d\theta = \frac{1}{2} \cdot \frac{1}{20} \cdot \int_0^{2\pi} d\theta = \frac{1}{2} \cdot \frac{1}{20} \cdot 2\pi = \frac{\pi}{20}.$$

4-72 ∎∎∎

Find the volume bounded above by the surface $z = x^2 - y^2$, below by the xy - plane, and laterally by the cylinder $x^2 + y^2 = 1$.

****

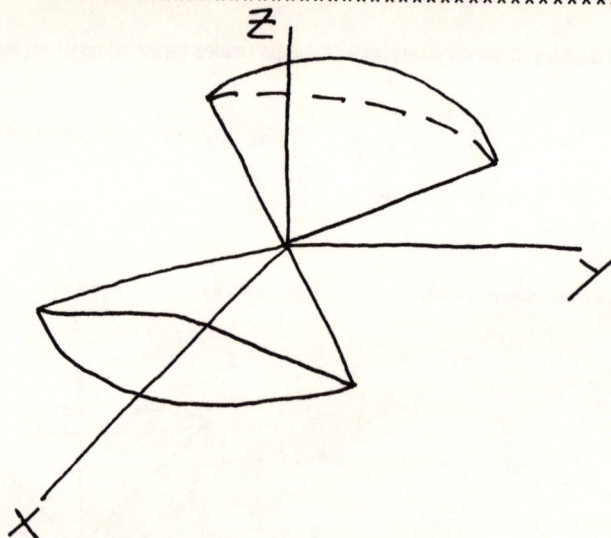

Use cylindrical coordinates and symmetry,

$$V = 4 \int_0^{\pi/4} \int_0^1 \int_0^{r^2\cos^2\theta - r^2\sin^2\theta} r \, dz \, dr \, d\theta$$

$$= 4 \int_0^{\pi/4} \int_0^1 r^3 \cos 2\theta \, dr \, d\theta$$

$$= 4 \int_0^{\pi/4} \frac{\cos 2\theta}{4} \, d\theta$$

$$= \frac{\sin 2\theta}{2} \Big|_0^{\pi/4} = \frac{1}{2}$$

━ 4-73

Find the volume of the region inside the cylinder $x^2 + y^2 = 7$ which is bounded below by the xy-plane and above by the sphere $x^2 + y^2 + z^2 = 16$.

****

USE CYLINDRICAL COORDINATES.

$x^2 + y^2 = 7$ BECOMES $r = \sqrt{7}$.

$x^2 + y^2 + z^2 = 16$ BECOMES $r^2 + z^2 = 16$

$$V = \int_0^{2\pi} d\theta \int_0^{\sqrt{7}} dr \int_0^{\sqrt{16-r^2}} r\, dz$$

$$= \int_0^{2\pi} d\theta \int_0^{\sqrt{7}} \sqrt{16-r^2}\; r\, dr$$

$$= -\frac{1}{2} \int_0^{2\pi} d\theta \int_0^{\sqrt{7}} (16-r^2)^{\frac{1}{2}} (-2r)\, dr$$

$$= -\frac{1}{2} \int_0^{2\pi} \frac{2}{3}(16-r^2)^{\frac{3}{2}} \Big|_0^{\sqrt{7}} d\theta$$

$$= -\frac{1}{3} \int_0^{2\pi} \left(9^{\frac{3}{2}} - 16^{\frac{3}{2}}\right) d\theta$$

$$= \frac{37}{3} \int_0^{2\pi} d\theta = \frac{74\pi}{3} \text{ CUBIC UNITS}$$

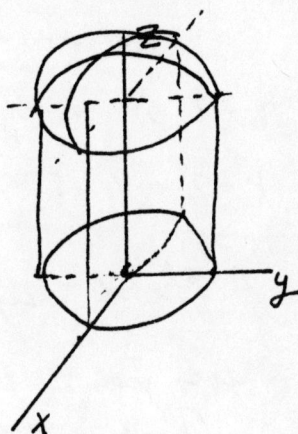

4-74 ▬▬▬▬▬▬▬▬▬▬▬▬▬▬▬▬▬▬▬▬▬▬▬▬▬▬▬▬▬▬▬▬▬▬▬▬▬▬

Let R be the three dimensional region in the first octant that is
outside the cylinder r = 1 and inside the sphere $r^2 + z^2 = 4$
(cylindical coordinates) Set up, but do not integrate, the iterated
(triple) integral for the volume of R. Integrate with respect to
z,r,θ, in that order.

****

$r = 1$ is a cylinder of radius 1; its axis is the z-axis

$r^2 + z^2 = 4$ is a sphere of radius 2 with center at
the origin

A sketch helps:

z goes from 0 to $\sqrt{4-r^2}$
(roof is sphere)

r goes from cylinder to
sphere
i.e. from r=1 to r=2

θ goes from positive
x-axis to positive
y-axis, i.e
from 0 to $\pi/2$

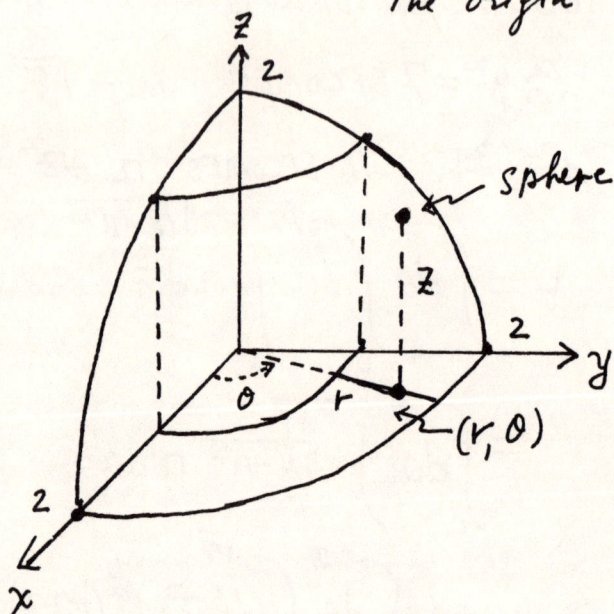

So, $Vol = \displaystyle\int_0^{\pi/2} \int_1^2 \int_0^{\sqrt{4-r^2}} r \, dz \, dr \, d\theta$

Note r in integrand for cylindrical coordinates

■■ **4-75**

Find the volume of the region above the paraboloid $z = x^2 + y^2$ and below

the hemisphere $z - 9 = \sqrt{9-x^2-y^2}$.
****

$$(z-9) = \sqrt{9-x^2-y^2}$$

$$\leftarrow z = x^2 + y^2$$

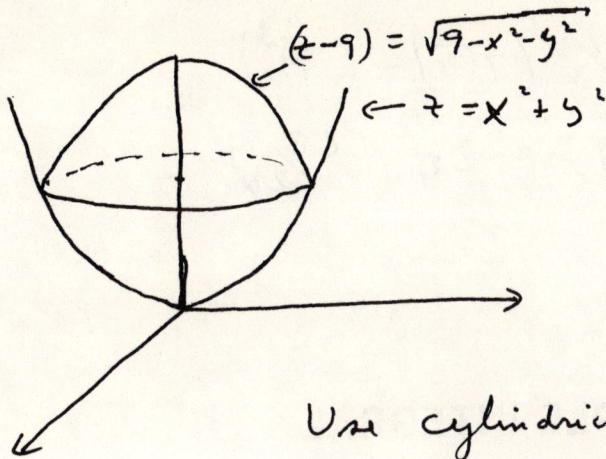

Use cylindrical coords. $\begin{cases} x = r\cos\theta \\ y = r\sin\theta \\ z = z. \end{cases}$

$$z = x^2 + y^2 = r^2$$

$$z - 9 = \sqrt{9 - x^2 - y^2}$$

$$\text{so } z = 9 + \sqrt{9 - r^2}$$

So $\text{vol} = \displaystyle\int_0^{2\pi} d\theta \int_0^3 (z_2 - z_1)\, r\, dr.$

$$= \int_0^{2\pi} d\theta \int_0^3 \left[(9 + \sqrt{9-r^2}) - r^2 \right] r\, dr.$$

$$= 2\pi \int_0^3 9r - r^3 + r(9 - r^2)^{\frac{1}{2}}\, dr$$

$$u = 9 - r^2$$
$$du = -2r\, dr$$

$$= 2\pi \left[\frac{9r^2}{2} - \frac{r^4}{4} \Big|_0^3 - \frac{1}{2} \int u^{\frac{1}{2}} du. \right]$$

$$= 2\pi \left[\frac{81}{2} - \frac{81}{4} - \frac{1}{2} \cdot \frac{2}{3} u^{\frac{3}{2}} \right]$$

$$= 2\pi \left(\frac{81}{2} - \frac{1}{3} \left(\sqrt{9-x^2} \right)^3 \Big|_0^3 \right)$$

$$= 2\pi \left(8\frac{1}{2} - \frac{1}{3} \left((\sqrt{0})^3 - (\sqrt{9})^3 \right) \right)$$

$$= 2\pi \left(8\frac{1}{2} - 9 \right) = 2\pi \left(6\frac{3}{2} \right)$$

$$= 63\pi$$

THE TRIPLE INTEGRAL
SPHERICAL COORDINATES

4-76 ■■

Use spherical coordinates to evaluate $\iiint_S (x^2 + y^2 + z^2)^2 \, dV$ where S is the solid in the first octant bounded by the sphere $x^2 + y^2 + z^2 = 4$.

****

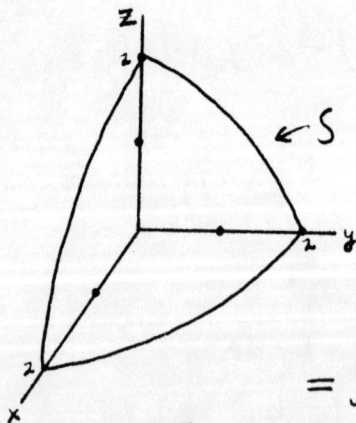

$$\iiint_S \underbrace{(x^2+y^2+z^2)^2}_{} \, dV$$

$$= \int_{\rho=0}^{\rho=2} \int_{\theta=0}^{\theta=\frac{\pi}{2}} \int_{\varphi=0}^{\varphi=\frac{\pi}{2}} \rho^4 \cdot \overbrace{\rho^2 \sin\varphi \, d\varphi \, d\theta \, d\rho}^{\text{Jacobian for sph. coords.}}$$

$$= \int_0^2 \int_0^{\frac{\pi}{2}} \int_0^{\frac{\pi}{2}} \rho^6 \sin\varphi \, d\varphi \, d\theta \, d\rho$$

$$= \int_0^2 \int_0^{\frac{\pi}{2}} \rho^6 [-\cos\varphi]_0^{\frac{\pi}{2}} \, d\theta \, d\rho = \int_0^2 \int_0^{\frac{\pi}{2}} \rho^6 \, d\theta \, d\rho$$

$$= \frac{\pi}{2} \int_0^2 \rho^6 \, d\rho = \frac{\pi}{2} \cdot \frac{\rho^7}{7} \Big|_0^2 = \frac{\pi}{2} \cdot \frac{2^7}{7} = \frac{64\pi}{7}.$$

■■ **4-77**

Evaluate the following integral by changing to spherical coordinates.

$$\int_{-5}^{5} \int_{0}^{\sqrt{25-x^2}} \int_{0}^{\sqrt{25-x^2-y^2}} \frac{1}{\sqrt{x^2+y^2+z^2}} \, dz \, dy \, dx$$

****

Note that the integration occurs over a region in the xy-plane and that

z goes from 0 to $\sqrt{25-x^2-y^2}$ (sphere of radius 5)

y goes from 0 to $\sqrt{25-x^2}$ (circle of radius 5)

x goes from -5 to 5

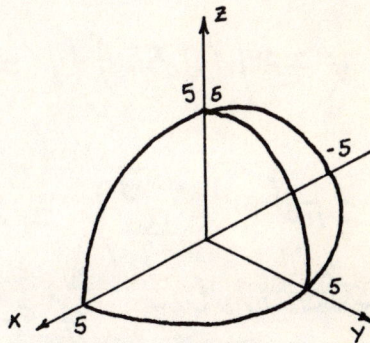

The spherical limits are: $0 \le \rho \le 5$, $0 \le \phi \le \pi/2$ and $0 \le \theta \le \pi$

In addition, $\dfrac{1}{\sqrt{x^2+y^2+z^2}} = \dfrac{1}{\rho}$

So we have $\displaystyle\int_{0}^{\pi} \int_{0}^{\pi/2} \int_{0}^{5} \frac{1}{\rho} \, \rho^2 \sin\phi \, d\rho \, d\phi \, d\theta$

$$= \frac{25}{2} \int_{0}^{\pi} \int_{0}^{\pi/2} \sin\phi \, d\phi \, d\theta$$

$$= \frac{25}{2} \int_{0}^{\pi} d\theta = \underline{\underline{\frac{25\pi}{2}}}$$

4-78 ■■■

Find the volume of a sphere using single, double and triple integrals.

****

$$x^2 + y^2 = a^2 \implies y = \pm\sqrt{a^2 - x^2}$$

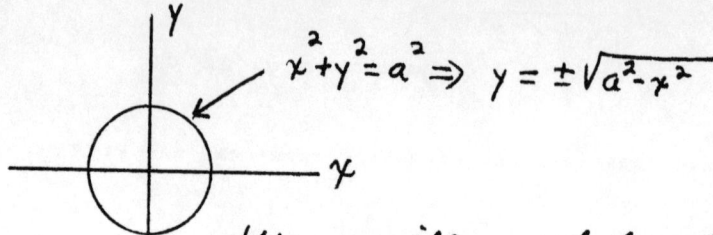

We will rotate the circle about the x-axis and use the disk method.

$$V = 2\pi \int_0^a \left(\sqrt{a^2 - x^2}\right)^2 dx = 2\pi \int_0^a (a^2 - x^2) dx = 2\pi \left\{ a^2 x - \frac{x^3}{3} \Big|_0^a \right\} =$$

$$2\pi \left\{ a^3 - \frac{a^3}{3} \right\} = 2\pi \left\{ \frac{2a^3}{3} \right\} = \frac{4a^3\pi}{3} \text{ which is}$$

the volume of a sphere of radius a.

$$x^2 + y^2 + z^2 = a^2 \implies z = \pm\sqrt{a^2 - x^2 - y^2}$$

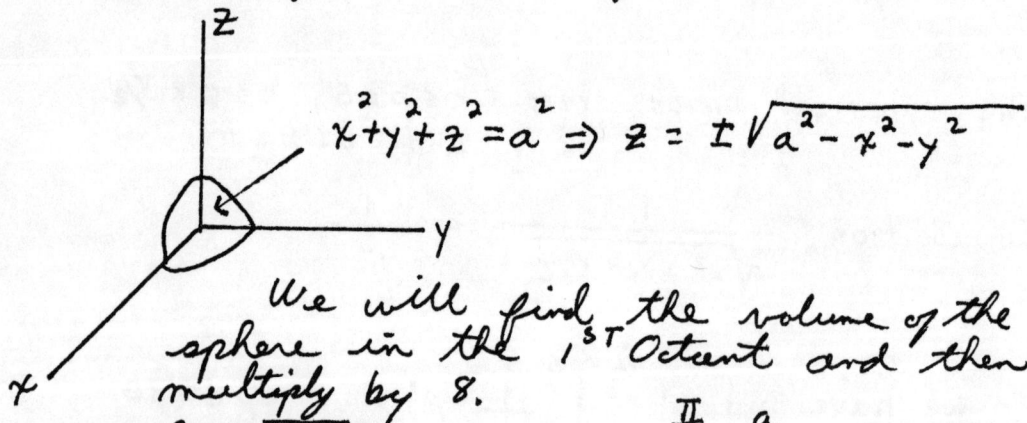

We will find the volume of the sphere in the 1ST Octant and then multiply by 8.

$$V = 8 \int_0^a \int_0^{\sqrt{a^2 - x^2}} \sqrt{a^2 - x^2 - y^2} \, dy \, dx = 8 \int_0^{\frac{\pi}{2}} \int_0^a \sqrt{a^2 - r^2} \, r \, dr \, d\theta =$$

$$-4 \int_0^{\frac{\pi}{2}} \int_0^a (a^2 - r^2)^{1/2} (-2r \, dr) d\theta = -4 \int_0^{\frac{\pi}{2}} \frac{2}{3} (a^2 - r^2)^{3/2} \Big|_0^a d\theta =$$

$$\frac{-8}{3} \int_0^{\frac{\pi}{2}} \left(\left(a^2 - a^2\right)^{3/2} - \left(a^2 - 0\right)^{3/2} \right) d\theta = \frac{8a^3}{3} \int_0^{\frac{\pi}{2}} d\theta =$$

$$\frac{8a^3}{3} \left\{ \theta \Big|_0^{\frac{\pi}{2}} \right\} = \frac{8a^3}{3} \left(\frac{\pi}{2} \right) = \frac{4a^3 \pi}{3}.$$

Now we will find the volume of the sphere in the 1st octant with spherical coordinates.

$$V = 8 \int_0^{\frac{\pi}{2}} \int_0^{\frac{\pi}{2}} \int_0^a \rho^2 \sin\phi \, d\rho \, d\phi \, d\theta = 8 \int_0^{\frac{\pi}{2}} \int_0^{\frac{\pi}{2}} \frac{\rho^3}{3} \sin\phi \Big|_0^a d\phi \, d\theta$$

$$= \frac{8a^3}{3} \int_0^{\frac{\pi}{2}} -\cos\phi \Big|_0^{\frac{\pi}{2}} d\theta = \frac{8a^3}{3} \int_0^{\frac{\pi}{2}} \left(-\cos\frac{\pi}{2} + \cos 0 \right) d\theta$$

$$= \frac{8a^3}{3} \int_0^{\frac{\pi}{2}} d\theta = \frac{8a^3}{3} \left\{ \theta \Big|_0^{\frac{\pi}{2}} \right\} = \frac{8a^3}{3} \left(\frac{\pi}{2} \right) = \frac{4a^3 \pi}{3}.$$

4-79 ▬▬▬▬▬▬▬▬▬▬▬▬▬▬▬▬▬▬▬▬▬▬▬▬

A region W in R^3 is described completely by $x \geq 0$, $y \geq 0$, $z \geq 0$, and $x^2 + y^2 + z^2 \leq 4$.
a) Describe or sketch this region.
b) Write an integral in rectangular coordinates which gives the volume of W. Do not work out this integral.
c) Write an integral in spherical coordinates which gives the volume of W. Find volume of W using this integral.

****

a) The first 3 inequalities place W in the first octant of R^3, while the last inequality places W inside the solid sphere of radius 2. So W looks like $\frac{1}{8}$ of a solid sphere of radius 2, as shown below. An interior slice is also shown for clarity.

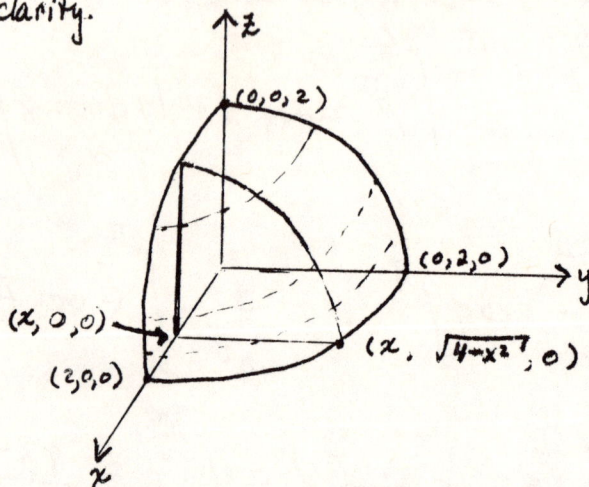

b) One possible answer is

$$\text{Vol of } W = \int_{x=0}^{2} \int_{y=0}^{\sqrt{4-x^2}} \int_{z=0}^{\sqrt{4-x^2-y^2}} dz\, dy\, dx$$

(This results from slicing the region in vertical slices which are perpendicular to the x-axis. A typical slice of this type has been drawn in the picture above. Other methods of slicing W will lead to other correct answers.)

c) Since W lies in the first octant, both θ and φ are restricted to lie in $[0, \frac{\pi}{2}]$, while ρ takes values in $[0, 2]$. The Jacobean for spherical coords is $\rho^2 \sin \varphi$.

$$\text{Vol } W = \int_{\rho=0}^{2} \int_{\theta=0}^{\pi/2} \int_{\varphi=0}^{\pi/2} \rho^2 \sin \varphi \, d\varphi \, d\theta \, d\rho$$

$$= \int_{\rho=0}^{2} \rho^2 \int_{\theta=0}^{\pi/2} \underbrace{-\cos\varphi \Big]_{0}^{\pi/2}}_{1} \, d\theta \, d\rho$$

$$= \int_{\rho=0}^{2} \rho^2 \; \theta \Big]_{0}^{\pi/2} \quad = \quad \frac{\pi}{2} \cdot \frac{\rho^3}{3} \Big]_{0}^{2} \; = \; \frac{8\pi}{6} \; = \; \boxed{\frac{4\pi}{3}}$$

4-80 ▪▪▪

Use a triple integral in spherical coordinates to find the volume of that part of the sphere $x^2 + y^2 + z^2 = 9$ which lies inside the cone $z = \sqrt{x^2 + y^2}$.

****

In spherical coord. :   sphere is $\rho = 3$

cone is $\phi = \frac{\pi}{4}$

$$\therefore \quad Vol = \int_0^{2\pi} \int_0^{\frac{\pi}{4}} \int_0^3 \rho^2 \sin\phi \; d\rho \; d\phi \; d\theta$$

$$= \int_0^{2\pi} \int_0^{\frac{\pi}{4}} \left(\frac{1}{3} \rho^3 \Big|_0^3 \right) \sin\phi \; d\phi \; d\theta$$

$$= \int_0^{2\pi} \int_0^{\frac{\pi}{4}} 9 \sin\phi \; d\phi \; d\theta$$

$$= \int_0^{2\pi} \left(-9 \cos\phi \Big|_0^{\frac{\pi}{4}} \right) d\theta = \int_0^{2\pi} \left(9 - \frac{9}{\sqrt{2}} \right) d\theta$$

$$= \left(9 - \frac{9}{\sqrt{2}} \right) \theta \Big|_0^{2\pi} = 18\pi \left(1 - \frac{1}{\sqrt{2}} \right)$$

■■ **4-81**

Find the mass of that portion of the solid bounded above by the sphere $x^2 + y^2 + z^2 = 3$ which lies in the first octant, if the density varies as the distance from the center of the sphere.

****

Use Spherical Coordinates.

Density $= K\rho$.

$$\text{Mass} = \int_0^{\frac{\pi}{2}} d\theta \int_0^{\frac{\pi}{2}} d\phi \int_0^{\sqrt{3}} K\rho^3 \sin\phi \, d\rho$$

$$= \int_0^{\frac{\pi}{2}} d\theta \int_0^{\frac{\pi}{2}} (K \sin\phi) \left.\frac{\rho^4}{4}\right|_0^{\sqrt{3}} d\phi$$

$$= \frac{9K}{4} \int_0^{\frac{\pi}{2}} \left[-\cos\phi\right]_0^{\frac{\pi}{2}} d\theta = \frac{9K}{4} \int_0^{\frac{\pi}{2}} d\theta = \frac{9K\pi}{8}$$

STOKES' THEOREM

■■ **4-82**

Evaluate the following line integral where the path C is the curve of intersection of the paraboloid $3z = x^2 + y^2$ with the plane $3x + 4y + z = 12$.

$$\oint_C (12x^2y^2 - 6yz^2)dx + (8x^3y - 6xz^2)dy - 12xyz \, dz$$

****

USE THE THEOREM OF STOKES.

$$\oint_C (12x^2y^2 - 6yz^2)dx + (8x^3y - 6xz^2)dy - 12xyz\,dz = \oint_C Pdx + Qdy + Rdz$$

$$= \int\int \left(\frac{\partial R}{\partial y} - \frac{\partial Q}{\partial z}\right)dy \wedge dz + \left(\frac{\partial P}{\partial z} - \frac{\partial R}{\partial x}\right)dz \wedge dx + \left(\frac{\partial Q}{\partial x} - \frac{\partial P}{\partial y}\right)dx \wedge dy$$

$$= \int\int \left(-12xz + 12xz\right)dy\,dz + \left(-12yz + 12yz\right)dz\,dx + \left(24x^3y - 6z^2 - 24x^2y + 6z^2\right)dx\,dy$$

$$= 0$$

4-83 ■■■

Evaluate the flux integral $\iint (2x\underline{i} - y\underline{j} + 3z\underline{k})$ $\underline{n}ds$ over the boundary of the ball $x^2 + y^2 + z^2 \leq 9$.

****

RECALL THE DIVERGENCE THM.

$$\iint_{\partial G} (\underline{F} \cdot \underline{n})\,dS = \iiint_G (\nabla \cdot \underline{F})\,dV.$$

$$\underline{\nabla} = \left(\frac{d}{dx}, \frac{d}{dy}, \frac{d}{dz}\right)$$

$$\underline{\nabla} \cdot \underline{F} = 2 - 1 + 3 = 4.$$

so

$$I = 4 \,(\text{vol ball}) = 4 \cdot \frac{4}{3}\pi r^3 = \frac{16}{3}\pi \cdot 27$$

$$= 144\pi$$

■■**4-84**

Let F be the vector function defined by

$$\vec{F}(x,y,z) = x^2 y^2 \vec{i} + x^2 z^2 \vec{j} + y^2 z^2 \vec{k}$$

Let C be the rectangular path from (1,1,2) to (3,1,2) to (3,5,2) to (1,5,2) to (1,1,2). Use Stokes' theorem to evaluate the line integral

$$\int_C \vec{F} \cdot \vec{T} \, ds \, ,$$

where $\vec{T}$ is the unit tangent vector to C.

****

Clearly, the unit normal to the surface enclosed by C is $\vec{k}$.

$$curl(\vec{F}) = \begin{vmatrix} \vec{i} & \vec{j} & \vec{k} \\ \frac{\partial}{\partial x} & \frac{\partial}{\partial y} & \frac{\partial}{\partial z} \\ x^2 y^2 & x^2 z^2 & y^2 z^2 \end{vmatrix} = (2yz^2 - 2x^2 z)\vec{i}$$
$$- (0-0)\vec{j}$$
$$+ (2xz^2 - 2x^2 y)\vec{k}$$

Stokes' thm $\Rightarrow$

$$\int_C \vec{F} \cdot \vec{T} \, ds = \int_1^3 \int_1^5 curl\, \vec{F} \cdot \vec{n} \, dy \, dx$$
$$\underset{= \vec{k}}{\uparrow}$$

$$= \int_1^3 \int_1^5 (2x(2^2) - 2x^2 y) \, dy \, dx = \int_1^3 8xy - x^2 y^2 \Big|_1^5 \, dx$$

$$= \int_1^3 (40x - 25x^2 - 8x + x^2) \, dx = \int_1^3 (32x - 24x^2) dx$$

$$= 16x^2 - 8x^3 \Big|_1^3 = 16(9-1) - 8(27-1)$$

$$= 128 - 208 = -80$$

4-85 ■■

Evaluate:

$$\iint_S \vec{F} \cdot \vec{n} \, dS$$

where S is the cube bounded by the planes x = ±1, y = ±1, z = ±1
$\vec{F} = x^2 y \vec{i} + xy \vec{j} + y^2 z^3 \vec{k}$ and n is the outward normal.

****

By Gauss' Divergence theorem

$$\iint_S \vec{F} \cdot \vec{n} \, dS = \iiint_R \operatorname{div} \vec{F} \, dx \, dy \, dz$$

thus in this case $\vec{F} = \langle x^2 y, \, xy, \, y^2 z^3 \rangle$

$$\operatorname{div} F = 2xy + x + 3y^2 z^2$$

and hence

$$\iint_S \vec{F} \cdot \vec{n} \, dS = \int_{-1}^{1} \int_{-1}^{1} \int_{-1}^{1} (2xy + x + 3y^2 z^2) \, dx \, dy \, dz$$

$$= \int_{-1}^{1} \int_{-1}^{1} \left[x^2 y + \frac{x^2}{2} + 3y^2 z^2 x \right] \Bigg|_{x=-1}^{1} dy \, dz$$

$$= \int_{-1}^{1} \int_{-1}^{1} 6y^2 z^2 \, dy \, dz$$

$$= \int_{-1}^{1} \left[2y^3 z^2 \right] \Bigg|_{y=-1}^{1} dz$$

$$= \int_{-1}^{1} 4z^2 \, dz$$

$$= \frac{4z^3}{3} \Bigg|_{-1}^{1} = \boxed{\frac{8}{3}}$$

━━━━━━━━━━━━━━━━━━━━━━━━━━━━━━━━━━━━━━**4-86**

Use Stokes' Theorem to evaluate $\int_S \vec{\nabla} \times \vec{F} \cdot \hat{n}\ dS$ where F = $4y\hat{i} - 2z\hat{j} + 5x\hat{k}$ and S is the surface area of the paraboloid $4 - z = x^2 + y^2$ above the x - y plane.

****

$$\int_S \vec{\nabla} \times \vec{F} \cdot \hat{n}\, d\vec{S} = \oint_C \vec{F} \cdot d\vec{r}$$

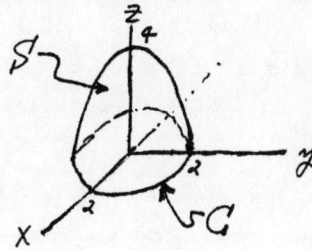

$$\vec{F} \cdot d\vec{r} = \left(4y\,\hat{i} - 2z\hat{j} + 5x\,\hat{k}\right) \cdot$$

$$\left(\hat{i}\,dx + \hat{j}\,dy + \hat{k}\,dz\right) = 4y\,dx - 2z\,dy + 5x\,dz$$

On C, $z = 0$ and $\vec{F} \cdot d\vec{r} = 4y\,dx - 0 + 0$

C is the circle $x^2 + y^2 = 4$; transform to polar

coordinates: $\quad x = r\cos\theta = 2\cos\theta$
$$y = r\sin\theta = 2\sin\theta$$

$$\oint_C \vec{F} \cdot d\vec{r} = \oint_C 4y\,dx = \int_0^{2\pi} 4 \cdot 2\sin\theta\, d(2\cos\theta)$$

$$= -16 \int_0^{2\pi} \sin\theta \cdot \sin\theta\, d\theta = -16 \int_0^{2\pi} \sin^2\theta\, d\theta$$

$$= -16 \int_0^{2\pi} \frac{1 - \cos 2\theta}{2}\, d\theta = -8\left(\theta - \frac{\sin 2\theta}{2}\right)\Big|_0^{2\pi}$$

$$= -8\left(2\pi - 0 - 0 + 0\right) = -16\pi$$

Therefore, $\quad \int_S \vec{\nabla} \times \vec{F} \cdot \hat{n}\, d S = -16\pi.$

4-87

Express Stoke's theorem, which is given as

$$\oint_C F \cdot dr = \iint_S (\text{curl } F) \cdot n \, dS,$$ in terms of P, Q, and R if they have

continuous first-order partial derivates throughout a region which contains
S and C. Find the curl of F if F (x,y,z)
$= yz \, i + z^2 x \, j + yz \, k.$

****

Stoke's theorem in terms of $P, Q,$ and R can be written as:

$$\oint_C (P \, dx + Q \, dy + R \, dz)$$

$$= \iint_S \left[\left(\frac{\partial R}{\partial y} - \frac{\partial Q}{\partial z} \right) dy \, dz - \left(\frac{\partial R}{\partial x} - \frac{\partial P}{\partial z} \right) dz \, dx \right.$$

$$\left. + \left(\frac{\partial Q}{\partial x} - \frac{\partial P}{\partial y} \right) dx \, dy \right]$$

If $F(x, y, z) = yz \, \hat{i} + z^2 x \, \hat{j} + yz \, \hat{k}$, we write:

$$\text{curl } F = \begin{vmatrix} \hat{i} & \hat{j} & \hat{k} \\ \frac{\partial}{\partial x} & \frac{\partial}{\partial y} & \frac{\partial}{\partial z} \\ yz & z^2 x & yz \end{vmatrix}$$

$$= (z^2 - z) \, i - (0 - y) \, j + (z^2 - z) \, k$$

$$= z(z-1) \, i + y \, j + z(z-1) \, k \quad \underline{Ans.}$$

MISCELLANEOUS PROBLEMS

■■ **4-88**

A sphere of radius k has a volume of $\frac{4}{3}\pi k^3$. Set up the iterated integrals in rectangular, cylindrical, and spherical coordinates to show this.

****

Rectangular $\quad x^2 + y^2 + z^2 = k^2$

$$V = \int_{-k}^{k} \int_{-\sqrt{k^2-x^2}}^{\sqrt{k^2-x^2}} \int_{-\sqrt{k^2-x^2-y^2}}^{\sqrt{k^2-x^2-y^2}} dz\, dy\, dx$$

Cylindrical $\quad r^2 + z^2 = k^2$

$$V = \int_{0}^{2\pi} \int_{0}^{k} \int_{-\sqrt{k^2-r^2}}^{\sqrt{k^2-r^2}} r\, dz\, dr\, d\theta$$

Spherical $\quad \rho = k$

$$V = \int_{0}^{2\pi} \int_{0}^{\pi} \int_{0}^{k} \rho^2 \sin\phi \; d\rho\, d\phi\, d\theta$$

4-89

Define the vector function $\vec{F}$ by:

$$\vec{F}(x,y,z) = (2x-z)\vec{i} + x^2 y\vec{j} + xz^2\vec{k}$$

Use the Divergence theorem to evaluate the surface integral $\iint_S \vec{F} \cdot \vec{n}\, dS$, where S is the surface enclosing the unit cube with $\vec{n}$ the outward-pointing unit normal.

****

$$\text{Div}(\vec{F}) = 2 + x^2 + 2xz$$

$$\iint_S \vec{F} \cdot \vec{n}\, dS = \int_0^1 \int_0^1 \int_0^1 (2 + x^2 + 2xz)\, dz\, dy\, dx$$

$$= \int_0^1 \int_0^1 \left[(2+x^2)z + xz^2 \right]_0^1 dy\, dx = \int_0^1 \int_0^1 (2 + x + x^2)\, dy\, dx$$

$$= \int_0^1 (2+x+x^2)\, y \Big|_0^1 dx = \int_0^1 (2 + x + x^2)\, dx$$

$$= 2x + \frac{x^2}{2} + \frac{x^3}{3} \Big|_0^1 = 2 + \frac{1}{2} + \frac{1}{3} = \frac{17}{6}$$

▬▬▬▬▬▬▬▬▬▬▬▬▬▬▬▬▬▬▬▬▬▬▬▬▬ **4-90**

Given that $\nabla f(x,y) = (3x^2 + y^2)i + (2xy - 3)j$, find:

a. $f(x,y)$

b. $\displaystyle\int_{(1,2)}^{(2,1)} (3x^2 + y^2)dx + (2xy - 3)dy$

c. $\displaystyle\oint_C (3x^2 + y^2)dy + (2xy - 3)dy$, where C is the curve given by
$$4x^2 + 9y^2 = 36.$$

****

a. Since $f_x(x,y) = 3x^2 + y^2$, we integrate with respect to x and get $f(x,y) = x^3 + xy^2 + g(y)$ (where $g(y)$ is constant with respect to x)

Then $f_y(x,y) = 2xy + g'(y)$.

From the given expression we know that $f_y(x,y) = 2xy - 3$,

so $\quad g'(y) = -3$

thus $\quad g(y) = -3y + c$.

Therefore $f(x,y) = x^3 + xy^2 - 3y + c$

b. Because the integrand is a gradient, this integral is independent of path. Thus we evaluate the potential function, found in part a to get the answer.

$$x^3 + xy^2 - 3y \Big|_{(1,2)}^{(2,1)} = (8 + 2 - 3) - (1 + 4 - 6) = 8$$

c. Green's Theorem is not needed here because the integrand is a gradient. Since the curve ends in the same place it begins, the answer is 0.

4-91

Evaluate the line integral $\int xy\,dx - x\,dy$ over the curve $y = 1 - x^2$ from the point $(1,0)$ to the point $(0,1)$.

****

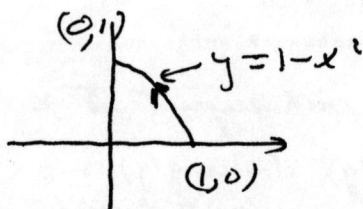

$$. y = 1 - x^2$$
$$x : 1 \longrightarrow 0.$$
$$dy = -2x\,dx$$

$$\int_C xy\,dx - x\,dy = \int_1^0 x(1-x^2)\,dx - x(-2x)\,dx.$$

$$= \int_1^0 (x + x^2)\,dx = \left. \frac{x^2}{2} + \frac{x^3}{3} \right|_1^0$$

$$= 0 - \left(\frac{1}{2} + \frac{1}{3}\right) = -\frac{5}{6}.$$

4-92

Let $M = 2xe^y + ye^x + 1$, $N = x^2e^y + e^x + y$. If possible, find a function φ for which $\vec{\nabla}\varphi = \langle M, N \rangle$.

****

$M_y = 2xe^y + e^x$, $N_x = 2xe^y + e^x$. Since $M_y \equiv N_x$, there does exist such φ. $\varphi_x = M \Rightarrow \varphi_x = 2xe^y + ye^x + 1 \Rightarrow$

$\varphi = x^2e^y + ye^x + x + g(y)$, $g(y)$ to be determined.

$\varphi_y = x^2e^y + e^x + g'(y)$, but also, $\varphi_y = N = x^2e^y + e^x + y$.

Thus $g'(y) = y$, $g(y) = \frac{y^2}{2} + C$. Setting $C = 0$, $g(y) = \frac{y^2}{2}$

and $\varphi = x^2e^y + ye^x + x + \frac{y^2}{2}$.

4-93

Find the work done by the force $\vec{F} = (2x + y)\vec{i} + (xy)\vec{j}$ in moving an object from $(1,0)$ to $(2,3)$ along the path C given by $x = t + 1$, $y = 3t$.

****

On C, $(1,0)$ is obtained when $t=0$ and $(2,3)$ when $t=1$,

$\text{work} = \int_C (2x+y)dx + xy\,dy = \int_0^1 [\overbrace{(2t+2+3t)}^{2x+y}\overbrace{dt}^{dx} + \overbrace{(t+1)3t}^{xy}\overbrace{3dt}^{dy}]$

$= \int_0^1 (14t + 9t^2 + 2)dt = 7t^2 + 3t^3 + 2t \Big|_0^1 = 12.$

4-94

Evaluate the line integral $\int_C xy\,dx + (x+y)\,dy$ where C is the curve $x = t^2$, $y = t^3$, $0 \le t \le 1$.

****

$$\int_C xy\,dx + (x+y)\,dy = \int_0^1 \left[\overbrace{t^2}^{x}\,\overbrace{t^3}^{y}\,\overbrace{2t\,dt}^{dx} + \overbrace{(t^2+t^3)}^{x+y}\,\overbrace{3t^2\,dt}^{dy} \right]$$

$$= \int_0^1 (2t^6 + 3t^4 + 3t^5)\,dt = \left. \frac{2t^7}{7} + \frac{3t^5}{5} + \frac{3t^6}{6} \right|_0^1 =$$

$$\frac{2}{7} + \frac{3}{5} + \frac{1}{2} = \frac{20 + 42 + 35}{70} = \frac{97}{70}.$$

4-95

Let $\varphi(x,y) = 3x^2y^3$. Note that $\nabla\varphi = \langle 6xy^3, 9x^2y^2 \rangle$. Evaluate the line integral $\int_C 6xy^3\,dx + 9x^2y^2\,dy$ where C is the curve $x = te^{t^3-1}$, $y = (2t+1)\cos(2\pi t)$, $0 \le t \le 1$.

****

The initial point of C ($t=0$) is $(0,1)$. The terminal point ($t=1$) is $(1,3)$. Since $\nabla\varphi = \langle 6xy^3, 9x^2y^2 \rangle$,

$$\int_C 6xy^3\,dx + 9x^2y^2 = \left. \varphi \right|_{(0,1)}^{(1,3)} = \left. 3x^2y^2 \right|_{(0,1)}^{(1,3)} = 81.$$